Significant Changes to the Seismic Load Provisions of ASCE 7-10

Other Titles of Interest

Minimum Design Loads for Buildings and Other Structures, ASCE/SEI 7-10. (ASCE Standard, 2010). Provides requirements for general structural design and includes means for determining dead, live, soil, flood, wind, snow, rain, atmospheric ice, and earthquake loads and their combinations that are suitable for inclusion in building codes and other documents. A detailed commentary of explanatory and supplementary information is included. (ISBN 978-0-7844-1085-1)

***Significant Changes to the Wind Load Provisions of ASCE 7-10 An Illustrated Guide*, by T. Eric Stafford, P.E.** (ASCE Press, 2010). Focuses on revisions to the wind load requirements set forth in the latest edition of the ASCE Standard for minimum design loads. (ISBN 978-0-7844-1116-2)

***Snow Loads: Guide to the Snow Load Provisions of ASCE 7-10*, by Michael O'Rourke.** Presents a detailed, authoritative interpretation of the snow load provisions of ASCE 7-10 by a respected engineering professional. (ISBN 978-0-7844-1111-7)

***Seismic Loads: Guide to the Seismic Load Provisions of ASCE 7-05*, by Finley A. Charney, Ph.D., P.E.** (ASCE Press, 2010). Presents a detailed, authoritative interpretation of the seismic load provisions of ASCE 7-05 by a respected engineering professional. (ISBN 978-0-7844-1076-9)

***Wind Loads: Guide to the Wind Load Provisions of ASCE 7-05*, by Kishor C. Mehta, Ph.D., P.E., and William Coulbourne, P.E.** (ASCE Press, 2010). Presents a detailed, authoritative interpretation of the wind load provisions of ASCE 7-05 by respected engineering professionals. (ISBN 978-0-7844-0858-2)

Significant Changes to the Seismic Load Provisions of ASCE 7-10

An Illustrated Guide

S. K. Ghosh, Ph.D.
Susan Dowty, S.E.
Prabuddha Dasgupta, Ph.D., P.E.

ASCE PRESS

Library of Congress Cataloging-in-Publication Data

Ghosh, S. K. (Satyendra Kumar)
Significant changes to the seismic load provisions of ASCE 7-10 an illustrated guide / S.K. Ghosh, Susan Dowty, Prabuddha Dasgupta.
p. cm.
Includes bibliographical references and index.
ISBN 978-0-7844-1117-9
1. Earthquake resistant design--Standards--United States. 2. Buildings--Standards--United States.
I. Dowty, Susan. II. Dasgupta, Prabuddha, 1975- III. American Society of Civil Engineers. IV. Title.
TA658.44.G49 2010
624.1'762021873--dc22
2010052404

Published by American Society of Civil Engineers
1801 Alexander Bell Drive
Reston, Virginia 20191
www.asce.org/pubs

ISBN 978-0-7844-1117-9
Manufactured in the United States of America.

18 17 16 15 14 13 12 11 10 1 2 3 4 5

TABLE OF CONTENTS

Part III- Chapter 12 Seismic Design Requirements for Building Structures

Part IV - Chapter 13 Seismic Design Requirements for Nonstructural Components

PREFACE

Significant Changes to the Seismic Load Provisions of ASCE 7-10 An Illustrated Guide is intended to familiarize structural engineers, architects, code officials, and others in the building construction and design industry with the changes to the seismic load requirements in the newest edition of *Minimum Design Loads for Buildings and Other Structures*, Standard ASCE/SEI 7-10. This reference book is organized into eight parts that generally follow the organization of the seismic chapters in ASCE 7-10. Each significant change is presented using the following format:

Section(s): *Here the sections affected by the significant change are listed.*

Section Title or Subject: *Here the subject of the significant change is identified.*

Change Type: *Here the significant change is described by one of the following or a combination thereof: modification, addition, deletion, relocation, clarification.*

At-a-Glance: *Here the meaning of the significant change is succinctly summarized.*

2010 Standard: *Here the significant change is shown in strike-out and underline format, where strike-out indicates text from ASCE 7-05 that has been deleted and underlining indicates new text added to ASCE 7-10.*

Analysis and Commentary: *Here the reason for the significant change is explained. Analysis, commentary, color photographs or other illustrations are used to enrich the reader's understanding.*

Although it was not possible to discuss each and every change to the seismic provisions, the ones that would be of most interest to or have significant impact on the industry are discussed in detail. This reference is intended to be a companion to ASCE 7-10; it is important to have the full set of seismic provisions in ASCE 7-10 on hand for reference when reading through this document. The commentary and opinions provided are those of the authors and do not necessarily represent the official position of ASCE.

ACKNOWLEDGEMENT

The authors wish to acknowledge their appreciation to T. Eric Stafford who got them involved in this project. His support, counsel, and patience through the duration of this project have been greatly valued.

The authors would also like to sincerely thank Phil Brazil for his in-depth and thorough review, which greatly enhanced the quality of this publication.

Part I
Terminology Changes Affecting Multiple Sections & Chapters

11.1.2, 11.2, 11.3, 12.2.3.3, Table 12.6-1, 12.8.1.3, 12.8.2.1, Table 12.12-1, 12.13.6, 12.14.1.1, 12.14.8.1, 12.14.4.2.1

Modification

Determination of Number of Stories

At a Glance

There are several requirements in ASCE 7 that are based on the number of stories in a structure. The determination of this number of stories is clarified.

2010 Standard

11.1.2 Scope

Every structure, and portion thereof, including nonstructural components, shall be designed and constructed to resist the effects of earthquake motions as prescribed by the seismic requirements of this standard. Certain non-building structures, as described in Chapter 15, are also within the scope and shall be designed and constructed in accordance with the requirements of Chapter 15. Requirements concerning alterations, additions, and change of use are set forth in Appendix 11B. Existing structures and alterations to existing structures need only comply with the seismic requirements of this standard where required by Appendix 11B. The following structures are exempt from the seismic requirements of this standard:

1. Detached one- and two-family dwellings that are located where the mapped, short period, spectral response acceleration parameter, S_s, is less than 0.4 or where the Seismic Design Category determined in accordance with Section 11.6 is A, B, or C.
2. Detached one- and two-family wood-frame dwellings not included in Exception 1 with not more than two stories above grade plane, satisfying the limitations of and constructed in accordance with the IRC.
3. *(no change to text)*
4. *(no change to text)*
5. *(no change to text)*

11.2 DEFINITIONS

The following definitions apply only to the seismic requirements of this standard.

~~**BASEMENT:** A basement is any story below the lowest story above grade.~~

GRADE PLANE: A horizontal reference plane representing the average of finished ground level adjoining the structure at all exterior walls. Where the finished ground level slopes away from the exterior walls, the grade ~~reference~~ plane is ~~shall be~~ established by the lowest points within the area between the structure ~~buildings~~ and the property ~~lot~~ line or, where the property ~~lot~~ line is more than 6 ft (1,829 mm) from the structure, between the structure and points 6 ft (1,829 mm) from the structure.

STORY: The portion of a structure between the tops of two successive ~~finished~~ floor surfaces and, for the topmost story, from the top of the floor surface ~~finish~~ to the top of the roof surface ~~structural element~~.

STORY ABOVE GRADE PLANE: A story in which the floor or roof surface at the top of the story is more than 6 ft (1,828 mm) above grade plane or is more than 12 ft (3,658 mm) above the finished ground level at any point on the perimeter of the structure. ~~Any story having its finished floor surface entirely above grade, except that a story shall be considered as a story above grade where the finished floor surface of the story immediately above is more than 6 ft (1,829 mm) above the grade plane, more than 6 ft (1,829 mm) above the finished ground level for more than 40 percent of the total structure perimeter, or more than 12 ft (3,658 mm) above the finished ground level at any point.~~

11.3 SYMBOLS

N = number of stories above the base (Section 12.8.2.1)

12.2.3.3 R, C_d, and Ω_0 Values for Horizontal Combinations

The value of the response modification coefficient, R, used for design in the direction under consideration shall not be greater than the least value of R for any of the systems utilized in that direction. The deflection amplification factor, C_d, and the overstrength factor, Ω_0, shall be consistent with R required in that direction.

EXCEPTION: Resisting elements are permitted to be designed using the least value of R for the different structural systems found in each independent line of resistance if the following three conditions are met: (1) Risk Category I or II building, (2) two stories or less above grade plane ~~in height~~, and (3) use of light-frame construction or flexible diaphragms. The value of R used for design of diaphragms in such structures shall not be greater than the least value for any of the systems utilized in that same direction.

TABLE 12.6-1 PERMITTED ANALYTICAL PROCEDURES

Seismic Design Category	Structural Characteristics	Equivalent Lateral Force Analysis, Section 12.8[a]	Modal Response Spectrum Analysis, Section 12.9[a]	Seismic Response History Procedures, Chapter 16[a]
B, C	All structures	P	P	P
D, E, F	Risk Category I or II buildings not exceeding 2 stories above the base ~~in height~~	P	P	P
	Structures of light frame construction	P	P	P
	Structures with no structural irregularities and not exceeding 160 ft in structural height	P	P	P
	Structures exceeding 160 ft in structural height with no structural irregularities and with $T < 3.5T_s$	P	P	P
	Structures not exceeding 160 ft in structural height and having only horizontal irregularities of Type 2, 3, 4, or 5 in Table 12.3-1 or vertical irregularities of Type 4, 5a, or 5b in Table 12.3-2	P	P	P
	All other structures	NP	P	P

[a]P : Permitted; NP: Not Permitted; $T_s = S_{D1}/S_{DS}$.

12.8.1.3 Maximum S_s Value in Determination of C_s

For regular structures five stories or less above the base as defined in Section 11.2 ~~in height~~ and with a period, T, of 0.5 s or less, C_s is permitted to be calculated using a value of 1.5 for S_S.

12.8.2.1 Approximate Fundamental Period

The approximate fundamental period (T_a), in s, shall be determined from the following equation:

$$T_a = C_t h_n^x \qquad (12.8\text{-}7)$$

h_n is the structural height as defined in Section 11.2 and the coefficients C_t and x are determined from Table 12.8-2.

Alternatively, it is permitted to determine the approximate fundamental period (T_a), in s, from the following equation for structures not exceeding 12 stories above the base as defined in Section 11.2 where ~~in height in which~~ the seismic force–resisting system consists entirely of concrete or steel moment resisting frames and the average story height is at least 10 ft (3 m):

$$T_a = 0.1N \qquad (12.8\text{-}8)$$

where N = number of stories above the base.

The approximate fundamental period, T_a, in s for masonry or concrete shear wall structures is permitted to be determined from Eq. 12.8-9 as follows:

$$T_a = \frac{0.0019}{\sqrt{C_w}} h_n \qquad (12.8\text{-}9)$$

where C_w is calculated from Eq. 12.8-10 as follows:

$$C_W = \frac{100}{A_B} \sum_{i=1}^{x} \left(\frac{h_n}{h_i} \right)^2 \frac{A_i}{\left[1 + 0.83 \left(\frac{h_i}{D_i} \right)^2 \right]} \qquad (12.8\text{-}10)$$

where

A_B = area of base of structure, ft^2
A_i = web area of shear wall i in ft^2
D_i = length of shear wall i in ft
h_i = height of shear wall i in ft
x = number of shear walls in the building effective in resisting lateral forces in the direction under consideration.

Table 12.12-1 Allowable Story Drift, Δ_a[a,b]

Structure	Risk Category		
	I or II	III	IV
Structures, other than masonry shear wall structures, 4 stories or less above the base as defined in Section 11.2, with interior walls, partitions, ceilings, and exterior wall systems that have been designed to accommodate the story drifts.	$0.025h_{sx}$[c]	$0.020h_{sx}$	$0.015h_{sx}$
Masonry cantilever shear wall structures[d]	$0.010h_{sx}$	$0.010h_{sx}$	$0.010h_{sx}$
Other masonry shear wall structures	$0.007h_{sx}$	$0.007h_{sx}$	$0.007h_{sx}$
All other structures	$0.020h_{sx}$	$0.015h_{sx}$	$0.010h_{sx}$

(no change to footnotes)

12.13.6 Requirements for Structures Assigned to Seismic Design Categories D through F

In addition to the requirements of Sections 11.8.2, 11.8.3, 14.1.8, and 14.2.3.2, the following foundation design requirements shall apply to structures assigned to Seismic Design Category D, E, or F. Design and construction of concrete foundation elements shall conform to the requirements of ACI 318, Section 21.8, except as modified by the requirements of this section.

EXCEPTION: Detached one- and two-family dwellings of light-frame construction not exceeding two stories ~~in height~~ above grade plane need only comply with the requirements for Sections 11.8.2, 11.8.3 (Items 2 through 4), 12.13.2, and 12.13.5.

12.14.1.1 Simplified Design Procedure

The procedures of this section are permitted to be used in lieu of other analytical procedures in Chapter 12 for the analysis and design of simple buildings with bearing wall or building frame systems, subject to all of the limitations listed in this section. Where these procedures are used, the seismic design category shall be determined from Table 11.6-1 using the value of S_{DS} from Section 12.14.8.1. The simplified design procedure is permitted to be used if the following limitations are met:

1. The structure shall qualify for Risk Category I or II in accordance with Table 1.5-1.
2. The site class, defined in Chapter 20, shall not be class E or F.
3. The structure shall not exceed three stories ~~in height~~ above grade plane.
4. The seismic force-resisting system shall be either a bearing wall system or building frame system,

as indicated in Table 12.14-1.

(no change in remainder of Section 12.14.1.1)

12.14.4.2.1 Horizontal Combination Different seismic force-resisting systems are permitted to be used in each of the two principal orthogonal building directions. Where a combination of different structural systems is utilized to resist lateral forces in the same direction, the value of R used for design in that direction shall not be greater than the least value of R for any of the systems utilized in that direction.

EXCEPTION: For buildings of light-frame construction or having flexible diaphragms and that are two stories or less ~~in height~~ <u>above grade plane</u>, resisting elements are permitted to be designed using the least value of R of the different seismic force–resisting systems found in each independent line of framing. The value of R used for design of diaphragms in such structures shall not be greater than the least value for any of the systems utilized in that same direction.

12.14.8.1 Seismic Base Shear

The seismic base shear, V, in a given direction shall be determined in accordance with Eq. 12.14-11:

$$V = \frac{FS_{DS}}{R} W \qquad (12.14\text{-}9)$$

where:

$$S_{DS} = \frac{2}{3} F_a S_S$$

where F_a is permitted to be taken as 1.0 for rock sites, 1.4 for soil sites, or determined in accordance with Section 11.4.3. For the purpose of this section, sites are permitted to be considered to be rock if there is no more than 10 feet (3m) of soil between the rock surface and the bottom of spread footing or mat foundation. In calculating S_{DS}, S_S shall be in accordance with Section 11.4.1, but need not be taken larger than 1.5.

F = 1.0 for ~~one-story~~ buildings <u>that are one story above grade plane</u>
F = 1.1 for ~~two-story~~ buildings <u>that are two stories above grade plane</u>
F = 1.2 for ~~three-story~~ buildings <u>that are three stories above grade plane</u>

(Remainder of section not affected by this change)

Analysis and Commentary

Many of the revisions to the wording in the definitions and provisions are for consistency and improvement in language.

The definition of "basement" is deleted because it serves no specific purpose in the seismic provisions of ASCE 7.

The defined term "STORY ABOVE GRADE" is changed to "STORY ABOVE GRADE PLANE" for consistency with the same term in the IBC and for internal consistency in ASCE 7 with the definition of "GRADE PLANE." The definition for STORY ABOVE GRADE PLANE now uses fewer words to essentially provide the same definition except that the second of three conditions is deleted for consistency with the IBC definition. Changing "floor surface of the story immediately above" to "floor or roof surface at the top of the story" corrects the definition so that it now specifically addresses the case where the story above grade plane is the topmost story of the building. Also, adding "on the perimeter of the structure" for the 12-foot limit establishes where the measurement takes places, whereas the previous definition was silent. See figure below for illustration of the definition STORY ABOVE GRADE PLANE.

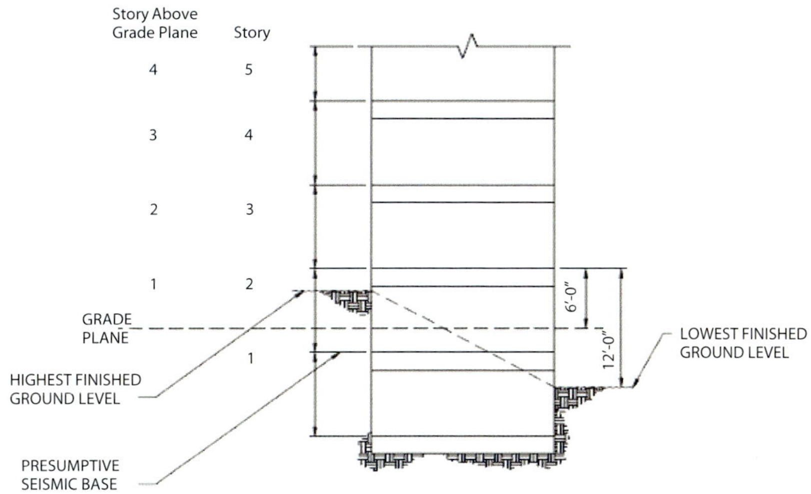

STORY 2 IS THE FIRST STORY ABOVE GRADE PLANE IF THE FLOOR (OR ROOF) SURFACE AT THE TOP OF THE STORY IS

1) 6'-0" ABOVE GRADE PLANE, OR
2) 12'-0" ABOVE THE FINISHED GROUND LEVEL AT ANY POINT ON THE PERIMETER OF THE STRUCTURE

There are several requirements in ASCE 7 that are based on the number of stories, which would include any basements in the structure. For example, a building with four stories above grade plane and four basements below grade plane is an eight-story building using the definition for STORY. Many ASCE 7 provisions are not based on including the basement stories in the determination of the number of stories. The language "above grade plane" or "above the base" is added to those ASCE 7 provisions where the intent is to determine the number of stories beginning at the first story above the grade plane or at the base (the level at which the horizontal seismic ground motions are considered to be imparted to the structure). In most buildings, the floor of the first story above grade plane is at the same approximate elevation as the base.

New commentary on the definition of Base in Section 11.2 is added to provide guidance on how to locate the base and to specifically address the following factors which affect the location of the seismic base:

1. Location of the grade relative to floor levels
2. Soil conditions adjacent to the building
3. Openings in basement walls
4. Location and stiffness of vertical elements of the seismic force-resisting system
5. Location and extent of seismic separations
6. Depth of basement
7. Manner in which basement walls are supported
8. Proximity to adjacent buildings
9. Slope of grade

Language Differences for Purpose of Determining Number of Stories

ASCE 7-05 Language	ASCE 7-10 Language
grade	grade plane
story above grade	story above grade plane
stories in height	stories above grade plane OR stories above the base OR stories above the base as defined in Section 11.2
stories in height above grade	stories above grade plane
stories	stories above grade plane OR stories above the base
finished floor surface	floor surface
floor finish	floor surface
roof structural element	roof surface
floor surface of the story immediately above	floor or roof surface at the top of the story

1.4, 11.2, 11.7.1, 12.1.1, 12.2.4, 12.4.4, 12.5.3, 12.13.6, 12.14.4.2.3, 12.14.7, 14.1.1, 14.2.1, 14.3.1, 14.4.1, 15.7.3, 15.7.6.1, 15.7.6.1.4, 15.7.10.5, 15.7.11.6, 16.2.1 and 16.2.4.2

Modification

Nonstructural Components, Structural Members, and Elements

At a Glance

A number of changes are made in various chapters so that the word "component" is consistently used to refer to nonstructural components and the term "member" or "element" is consistently used to refer to what was previously often described as structural component or simply component.

2010 Standard

11.2 DEFINITIONS

ATTACHMENTS: Means by which nonstructural components ~~and their supports~~ are secured or connected to the seismic force-resisting system of the structure. Such attachments include anchor bolts, welded connections, and mechanical fasteners.

COMPONENT: A part ~~or element~~ of an architectural, electrical, or mechanical~~, or structural~~ system.

Component, Nonstructural ~~Equipment~~: A part of an architectural, mechanical, or electrical system within or without a building or nonbuilding structure.

Component, Flexible: Nonstructural ~~C~~component having a fundamental period greater than 0.06 s.

Component, Rigid: Nonstructural ~~C~~component having a fundamental period less than or equal to 0.06 s.

PILE: Deep foundation ~~components including~~ element, which includes piers, caissons, and piles.

SHEAR PANEL: A floor, roof, or wall ~~component~~ element sheathed to act as a shear wall or diaphragm.

Because this code change revised numerous revisions in various places within Chapters 1, 11, 12, 14, 15 and 16, the entire code text is too lengthy to be included here.

Analysis and Commentary

The provisions of Chapter 13 apply to nonstructural components; however, the word "nonstructural" is frequently omitted and only the word "component" is used to mean "nonstructural component". On the other hand, in Chapters 11, 12, and 14-16, which apply to buildings and other structures, there are several references to (structural) components where (structural) members or elements are intended to be referred to. This change revises the definition of "component" to mean or refer to nonstructural components only and in Chapters 11, 12, and 14-16, the word "component" is changed to either member or element as appropriate. There are instances, however, where the use of the word "component" was not changed in Chapters 11, 12, and 14-16 because the continued use of the word "component" was judged to be appropriate.

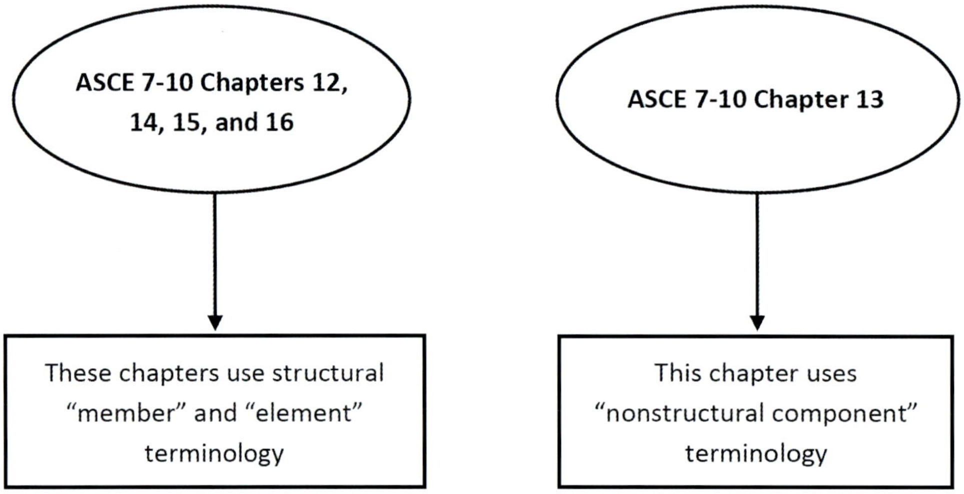

11.2, 11.3, Table 12.2-1, 12.2.5.4, 12.2.5.6.1, 12.2.5.6.2, 12.2.5.7.1, 12.2.5.7.2, 12.2.5.7.3, 12.3.3.2, 12.8.2.1, Table 12.14-1, 12.14.8.5, 15.4.1, Table 15.4-1, Table 15.4-2, 15.5.3.8, 15.7.10.6, 17.4.1, 18.1.3, 18.5.2.3, 19.2.1.1

Addition & Modification

Structural Height, h_n

At a Glance

A definition and corresponding notation are added for STRUCTURAL HEIGHT and the term "structural height" is used consistently throughout the provisions, rather than "building height" or "height."

2010 Standard

11.2 DEFINITIONS

STRUCTURAL HEIGHT: The vertical distance from the base to the highest level of the seismic force-resisting system of the structure. For pitched or sloped roofs, the structural height is from the base to the average height of the roof.

11.3 SYMBOLS

The unit dimensions used with the items covered by the symbols shall be consistent throughout except where specifically noted. Notation presented in this section applies only to the seismic requirements in this standard as indicated.

h_i, ~~h_n~~, h_x = the height above the base to Level i ~~n~~ or x, respectively

h_n = structural height as defined in Section 11.2.

Table 12.2-1 Design Coefficients and Factors for Seismic Force-Resisting Systems

Seismic Force-Resisting System	ASCE 7 Section Where Detailing Requirements Are Specified	Response Modification Coefficient, R [a]	Overstrength Factor, Ω_0 [g]	Deflection Amplification Factor, C_d [b]	Structural System Limitations Including ~~Building~~ Structural Height, h_n, (ft) Limits [c]

(no change to table entries)

a. Response modification coefficient, R, for use throughout the standard. Note R reduces forces to a strength level, not an allowable stress level.

b. Deflection amplification factor, C_d, for use in Sections 12.8.6, 12.8.7, and 12.9.2

c. NL = Not Limited and NP = Not Permitted. For metric units use 30.5 m for 100 ft and use 48.8 m for 160 ft. ~~Heights are measured from the base of the structure as defined in Section 11.2.~~

d. See Section 12.2.5.4 for a description of seismic force-resisting systems limited to buildings with a structural height, h_n, of 240 ft (73.2 m) or less.

e. See Section 12.2.5.4 for seismic force-resisting systems limited to buildings with a structural height, h_n, of 160 ft (48.8 m) or less.

(no change to footnotes f. through i.)

j. Steel ordinary concentrically braced frames are permitted in single-story buildings up to a structural height, h_n, of 60 ft (18.3 m) where the dead load of the roof does not exceed 20 psf (0.96 kN/m^2) and in penthouse structures.

k. An increase in the structural height, h_n, to 45 ft (13.7 m) is permitted for single story storage warehouse facilities.

(no change to footnotes l. through p.)

12.2.5.4 Increased ~~Building~~ Structural Height Limit for Steel Eccentrically Braced Frames, Steel Special Concentrically Braced Frames, Steel Buckling-restrained Braced Frames, Steel Special Plate Shear Walls and Special Reinforced Concrete Shear Walls.

The ~~height~~ limits on structural height, h_n, in Table 12-1 are permitted to be increased from 160 ft (50 m) to 240 ft (75 m) for structures assigned to Seismic Design Categories D or E and from 100 ft (30 m) to 160 ft (50 m) for structures assigned to Seismic Design Category F provided the seismic force-resisting systems are limited to steel eccentrically braced frames, steel special concentrically braced frames, steel buckling-restrained braced frames, steel special plate shear walls, or special reinforced concrete cast-in-place shear walls and both of the following requirements are met:
(no change in remainder of Section 12.2.5.4)

12.2.5.6 Steel Ordinary Moment Frames

12.2.5.6.1 Seismic Design Category D or E.

a. Single-story steel ordinary moment frames in structures assigned to Seismic Design Category D or E are permitted up to a structural height, h_n, of 65 ft (20 m) where the dead load supported by and tributary to the roof does not exceed 20 psf (0.96 kN/m^2). In addition, the dead load of the exterior walls more than 35 ft (10.6 m) above the base tributary to the moment frames shall not exceed 20 psf (0.96 kN/m^2).

EXCEPTION: *(no change in text)*

b. Steel ordinary moment frames in structures assigned to Seismic Design Category D or E not meeting the limitations set forth in Section 12.2.5.6.1.a are permitted within light-frame construction up to a structural height, h_n, of 35 ft (10.6 m) where neither the roof dead load nor the dead load of any floor above the base supported by and tributary to the moment frames exceeds 35 psf (1.68 kN/m^2). In addition, the dead load of the exterior walls tributary to the moment frames shall not exceed 20 psf (0.96 kN/m^2).

12.2.5.6.2 Seismic Design Category F. Single-story steel ordinary moment frames in structures assigned to Seismic Design Category F are permitted up to a structural height, h_n, of 65 ft (20 m) where the dead load supported by and tributary to the roof does not exceed 20 psf (0.96 kN/m^2). In addition, the dead load of the exterior walls tributary to the moment frames shall not exceed 20 psf (0.96 kN/m^2).

12.2.5.7 Steel Intermediate Moment Frames

12.2.5.7.1 Seismic Design Category D

a. Single-story steel intermediate moment frames in structures assigned to Seismic Design Category D are permitted up to a structural height, h_n, of 65 ft (20 m) where the dead load supported by and tributary to the roof does not exceed 20 psf (0.96 kN/m^2). In addition, the dead load of the exterior walls more than 35 ft (10.6 m) above the base tributary to the moment frames shall not exceed 20 psf (0.96 kN/m^2).

EXCEPTION: *(no change to text)*

b. Steel intermediate moment frames in structures assigned to Seismic Design Category D not meeting the limitations set forth in Section 12.2.5.7.1.a are permitted up to a structural height, , h_n, of 35 ft (10.6 m).

12.2.5.7.2 Seismic Design Category E

a. Single-story steel intermediate moment frames in structures assigned to Seismic Design Category E are permitted up to a structural height, h_n, of 65 ft (20 m) where the dead load supported by and tributary to the roof does not exceed 20 psf (0.96 kN/m^2). In addition, the dead load of the exterior walls more than 35 ft (10.6 m) above the base tributary to the moment frames shall not exceed 20 psf (0.96 kN/m^2).

EXCEPTION: *(no change to text)*

b. Steel intermediate moment frames in structures assigned to Seismic Design Category E not meeting the limitations set forth in Section 12.2.5.7.2.a are permitted up to a structural height, , h_n, of 35 ft (10.6 m) where neither the roof dead load nor the dead load of any floor above the base supported by and tributary to the moment frames exceeds 35 psf (1.68 kN/m^2). In addition, the dead load of the exterior walls tributary to the moment frames shall not exceed 20 psf (0.96 kN/m^2).

12.2.5.7.3 Seismic Design Category F

a. Single-story steel intermediate moment frames in structures assigned to Seismic Design Category F are permitted up to a structural height, h_n, of 65 ft (20 m) where the dead load supported by and tributary to the roof does not exceed 20 psf (0.96 kN/m^2). In addition, the dead loads of the exterior walls tributary to the moment frames shall not exceed 20 psf (0.96 kN/m^2).

b. Steel intermediate moment frames in structures assigned to Seismic Design Category F not meeting the limitations set forth in Section 12.2.5.7.3.a are permitted within light-frame construction up to a structural height, h_n, of 35 ft (10.6 m) where neither the roof dead load nor the dead load of any floor above the base supported by and tributary to the moment frames exceeds 35 psf (1.68 kN/m^2). In addition, the dead load of the exterior walls tributary to the moment frames shall not exceed 20 psf (0.96 kN/m^2).

12.3.3.2 Extreme Weak Stories

Structures with a vertical irregularity Type 5b as defined in Table 12.3-2, shall not be over two stories or 30 ft (9 m) in structural height, h_n.

EXCEPTION: The limit does not apply where the "weak" story is capable of resisting a total seismic force equal to Ω_0 times the design force prescribed in Section 12.8.

12.8.2.1 Approximate Fundamental Period

The approximate fundamental period (T_a), in s, shall be determined from the following equation:

$$T_a = C_t h_n^x \qquad (12.8\text{-}7)$$

where h_n is the structural height ~~in ft above the base to the highest level of the structure~~ as defined in Section 11.2 and the coefficients C_t and x are determined from Table 12.8-2.

Alternatively, it is permitted to determine the approximate fundamental period (T_a), in s, from the following equation for structures not exceeding 12 stories above the base as defined in Section 11.2 where the seismic force–resisting system consists entirely of concrete or steel moment resisting frames and the average story height is at least 10 ft (3 m):

$$T_a = 0.1\,N \qquad (12.8\text{-}8)$$

where N = number of stories above the base.

The approximate fundamental period, T_a, in s for masonry or concrete shear wall structures is permitted to be determined from Eq. 12.8-9 as follows:

$$T_a = 0.0019\,h_n / \sqrt{C_w} \qquad (12.8\text{-}9)$$

where ~~h_n is as defined in the preceding text and~~ C_w is calculated from Eq. 12.8-10 as follows:

(no change in remainder of Section 12.8.2.1)

Footnote d to Table 12.14-1 Design Coefficients and Factors For Seismic Force–Resisting Systems for Simplified Design Procedure

d. Light-frame walls with shear panels of all other materials are permitted up to 35 ft (10.6 m) in structural height, h_n, in Seismic Design Category D and are not permitted in Seismic Design Category E.

12.14.8.5 Drift Limits and Building Separation

Structural drift need not be calculated. Where a drift value is needed for use in material standards, to determine structural separations between buildings or from property lines, for design of cladding, or for other design requirements, it shall be taken as 1 percent of ~~building~~ structural height, h_n, unless computed to be less. All portions of the structure shall be designed to act as an integral unit in resisting seismic forces unless separated structurally by a distance sufficient to avoid damaging contact under the total deflection.

15.4.1 Design Basis

Nonbuilding structures having specific seismic design criteria established in reference documents shall be designed using the standards as amended herein. Where reference documents are not cited herein, nonbuilding structures shall be designed in compliance with Sections 15.5 and 15.6 to resist minimum seismic lateral forces that are not less than the requirements of Section 12.8 with the following additions and exceptions:

1. The seismic force-resisting system shall be selected as follows:

a. For nonbuilding structures similar to buildings, a system shall be selected from among the types indicated in Table 12.2-1 or Table 15.4-1 subject to the system limitations and the ~~height~~ limits on structural height, h_n, based on the seismic design category indicated in the table. The appropriate values of R, Ω_0, and C_d indicated in the selected table shall be used in determining the base shear, element design forces, and design story drift as indicated in this standard. Design and detailing requirements shall comply with the sections referenced in the selected table.

b. For nonbuilding structures not similar to buildings, a system shall be selected from among the types indicated in Table 15.4-2 subject to the system limitations and the ~~height~~ limits on structural height, h_n, based on seismic design category indicated in the table. The appropriate values of R, Ω_0, and C_d indicated in Table 15.4-2 shall be used in determining the base shear, element design forces, and design story drift as indicated in this standard. Design and detailing requirements shall comply with the sections referenced in Table 15.4-2.

(no change in remainder of Section 15.4.1)

Table 15.4-1 Seismic Coefficients For Nonbuilding Structures Similar To Buildings

Nonbuilding Structure Type	Detailing Requirements	R	Ω_0	C_d	Structural System and Structural Height, h_n, Limits (ft)[a]				
					B	C	D	E	F

[a] NL = no limit and NP = not permitted. ~~Height shall be measured from the base.~~

[b] Steel ordinary braced frames are permitted in pipe racks up to 65 ft (20 m).

[c] Steel ordinary moment frames and intermediate moment frames are permitted in pipe racks up to a height of 65 ft (20 m) where the moment joints of field connections are constructed of bolted end plates.

[d] Steel ordinary moment frames and intermediate moment frames are permitted in pipe racks up to a height of 35 ft (11 m).

~~[e] For the purpose of height limit determination, the height of the structure shall be taken as the height to the top of the structural frame making up the primary seismic force–resisting system.~~

Table 15.4-2 Seismic Coefficients For Nonbuilding Structures Not Similar To Buildings

Nonbuilding Structure Type	Detailing Requirements[c]	R	Ω_0	C_d	Structural Height, h_n, Limits (ft)[a, d]				
					B	C	D	E	F

(no changes to table entries)

[a] NL = no limit and NP = not permitted. ~~Height shall be measured from the base.~~

15.5.3.8 Seismic Displacements

Steel storage rack installations shall accommodate the seismic displacement of the storage racks and their contents relative to all adjacent or attached components and elements. The assumed total relative displacement for storage racks shall be not less than 5 percent of the structural height above the base, h_n, unless a smaller value is justified by test data or analysis in accordance with Section 11.1.4.

15.7.10.6 Welded Steel Water Storage Structures

Welded steel elevated water storage structures shall be designed and detailed in accordance with the seismic requirements of AWWA D100 with the structural height limits imposed by Table 15.4-2.

17.4.1 Equivalent Lateral Force Procedure

The equivalent lateral force procedure of Section 17.5 is permitted to be used for design of a seismically isolated structure provided that:

1. The structure is located at a Site with S_1 less than 0.60g.
2. The structure is located on a Site Class A, B, C, or D.
3. The structure above the isolation interface is less than or equal to four stories or 65 ft (19.8 m) in structural height, $\underline{h_n}$, measured from the base as defined in Section 11.2.

(no change in remainder of Section 17.4.1)

18.1.3 Notation

The following notations apply to the provisions of this chapter.

~~h_r = height of the structure above the base to the roof level, Section 18.5.2.3~~

18.5.2.3 Fundamental Mode Properties*

The fundamental mode shape, φ_{i1}, and participation factor, Γ_1, shall be determined by either dynamic analysis using the elastic structural properties and deformational characteristics of the resisting elements or using Eqs. 18.5-3 and 18.5-4:

$$\varphi_{i1} = h_i / ~~h_r~~ \underline{h_n} \quad (18.5\text{-}3)$$

(no change to Equation 18.5-4) (18.5-4)

where

h_i = the height ~~of the structure~~ above the base to Level i

~~h_r~~ $\underline{h_n}$ = the structural height ~~of the structure above the base to the roof level~~ as defined in Section 11.2

w_i = the portion of the total effective seismic weight, W, located at or assigned to Level i

(no change in remainder of Section 18.5.2.3)

***See ASCE 7-10 Errata for corrected text**

19.2.1.1 Effective Building Period.

The effective period (T) shall be determined as follows:

(no change to Equation 19.2-3) (19.2-3)

where

T = the fundamental period of the structure as determined in Section 12.8.2

$\bar{k}$ = the stiffness of the structure where fixed at the base, defined by the following:

(no change to Equation 19.2-4) (19.2-4)

where

$\bar{h}$ = the effective height of the structure, which shall be taken as 0.7 times the ~~total~~ structural height (h_n), except for structures where the gravity load is effectively concentrated at a single level, the effective height of the structure shall be taken as the height to that level

(no change in remainder of Section 19.2.1.1)

Analysis and Commentary

References were made to the height of the structure (e.g., building height, height, etc.) in several provisions of ASCE 7-05 but it was not clear how to measure the height of the structure. ASCE 7-05 Footnote (c) of Table 12.2-1 stated that it was to be measured from the base of the structure but was silent on where to complete the measurement above the base. ASCE 7-05 Footnote (a) of Tables 15.4-1 and 15-4.2 made a similar statement. Other ASCE 7-05 provisions were typically silent on how to measure it, whether from the base or to a point further up the structure. ASCE 7-05 Footnote (e) in Table 15.4-1 and Footnote (d) in Table 15.4-2 stated that it was to be measured to the top of the structural frame making up the primary seismic force-resisting system. These different ways of addressing the height of a structure are shown in the table below.

A new definition is added for "structural height" which provides clear direction on how to make this measurement. With the addition of the definition to Section 11.2, many revisions are made throughout the provisions for consistency and to eliminate confusion regarding how to measure the structural height.

In addition, the symbol "h_n" is assigned to the term structural height so that h_n is no longer available for general purposes. A symbol for "h_n" as the structural height is added to Section 11.3.

The "highest level of the seismic force-resisting system" rather than the roof or roof structural element is specified in the definition of "structural height" to prevent measurement of the structural height to the top of a penthouse or other rooftop structure, which are typically designed as separate seismic force-resisting systems or as nonstructural components (see Table 13.5-1). Where a penthouse is included in the seismic force-resisting system of the building structure, structural height would typically be measured to the roof of the penthouse due to the presence of a horizontal level of the seismic force-resisting system.

Language Differences for Height of Structure

<table>
<tr><th>ASCE 7-05 Language</th><th>ASCE 7-10 Language</th></tr>
<tr><td>building height
height
height of the structure above the base
~~height in ft above the base to the highest level of the structure~~</td><td>structural height or
structural height as defined in Section 11.2.</td></tr>
<tr><td>height limits</td><td>limits on structural height</td></tr>
<tr><td>Table 12.2-1 footnote c: Heights are measured from the base of the structure as defined in Section 11.2.</td><td rowspan="4">STRUCTURAL HEIGHT: The vertical distance from the base of the structure to the highest level of the seismic force-resisting system. For pitched or sloped roofs, the structural height is the average height of the roof.

and

h_n = structural height as defined in Section 11.2.</td></tr>
<tr><td>Tables 15.4-1 footnote a and 15.4-2 footnote a: Height shall be measured from the base.</td></tr>
<tr><td>Tables 15.4-1 footnote e and 15.4-2 footnote d: For the purpose of height limit determination, the height of the structure shall be taken as the height to the top of the structural frame making up the primary seismic force–resisting system.</td></tr>
<tr><td>Section 18.5.2.3:
h_r = height of the structure above the base to the roof level</td></tr>
</table>

12.2.1, Table 12.2-1, 12.4.1, 12.4.3, 12.4.3.2, 12.14.3, 12.14.3.2, 12.14.3.2.2, 18.6.3, 18.7.1.3, 18.7.2.3

Modification

Overstrength Factor

At a Glance

This change brings consistency in the use of the term "overstrength factor" throughout ASCE 7-10.

2010 Standard

12.2.1 Selection and Limitations

The basic lateral and vertical seismic force-resisting system shall conform to one of the types indicated in Table 12.2-1 or a combination of systems as permitted in Sections 12.2.2, 12.2.3, and 12.2.4. Each type is subdivided by the types of vertical elements used to resist lateral seismic forces. The structural system used shall be in accordance with the structural system limitations and the limits on structural height, h_n, contained in Table 12.2-1. The appropriate response modification coefficient, R, ~~system~~ overstrength factor, Ω_0, and the deflection amplification factor, C_d, indicated in Table 12.2-1 shall be used in determining the base shear, element design forces, and design story drift.

Each selected seismic force-resisting system shall be designed and detailed in accordance with the specific requirements for the system as set forth in the applicable reference document listed in Table 12.2-1 and the additional requirements set forth in Chapter 14.

Seismic force-resisting systems not contained in Table 12.2-1 are permitted provided analytical and test data are submitted to the authority having jurisdiction for approval that establish their dynamic characteristics and demonstrate their lateral force resistance and energy dissipation capacity to be equivalent to the structural systems listed in Table 12.2-1 for equivalent values of response modification coefficient, R, ~~system~~ overstrength factor ~~coefficient~~, Ω_0, and deflection amplification factor, C_d.

Table 12.2-1 Design Coefficients And Factors For Seismic Force-Resisting Systems

Seismic Force-Resisting System	ASCE 7 Section Where Detailing Requirements Are Specified	Response Modification Coefficient, R[a]	~~System~~ Overstrength Factor, Ω_0[g]	Deflection Amplification Factor, C_d[b]	Structural System Limitations Including Structural Height, h_n (ft) Limits[c]

12.4.1 Applicability

All members of the structure, including those not part of the seismic force-resisting system, shall be designed using the seismic load effects of Section 12.4 unless otherwise exempted by this standard. Seismic load effects are the axial, shear, and flexural member forces resulting from application of horizontal and vertical seismic forces as set forth in Section 12.4.2. Where specifically required, seismic load effects shall be modified to account for ~~system~~ overstrength, as set forth in Section 12.4.3.

12.4.3 Seismic Load Effect Including Overstrength Factor

Where specifically required, conditions requiring overstrength factor applications shall be determined in accordance with the following:

1. For use in load combination 5 in Section 2.3.2 or load combinations 5 and 6 in Section 2.4.1, E shall be taken equal to E_m as determined in accordance with Eq. 12.4-5 as follows:

$$E_m = E_{mh} + E_v \tag{12.4-5}$$

2. For use in load combination 7 in Section 2.3.2 or load combination 8 in Section 2.4.1, E shall be taken equal to E_m as determined in accordance with Eq. 12.4-6 as follows:

$$E_m = E_{mh} - E_v \tag{12.4-6}$$

where

E_m = seismic load effect including overstrength factor

E_{mh} = effect of horizontal seismic forces including ~~structural~~ overstrength factor as defined in Section 12.4.3.1

E_v = vertical seismic load effect as defined in Section 12.4.2.2

12.4.3.2 Load Combinations with Overstrength Factor

Where the seismic load effect with overstrength factor, E_m, defined in Section 12.4.3, is combined with the effects of other loads as set forth in Chapter 2, the following seismic load combination for structures not subject to flood or atmospheric ice loads shall be used in lieu of the seismic load combinations in either Section 2.3.2 or 2.4.1:
(no change in remainder of Section 12.4.3.2)

12.14.3 Seismic Load Effects and Combinations

All members of the structure, including those not part of the seismic force-resisting system, shall be designed using the seismic load effects of Section 12.14.3 unless otherwise exempted by this standard. Seismic load effects are the axial, shear, and flexural member forces resulting from application of horizontal and vertical seismic forces as set forth in Section 12.14.3.1. Where specifically required, seismic load effects shall be modified to account for ~~system~~ overstrength, as set forth in Section 12.14.3.2 ~~12.14.3.1.3~~.

12.14.3.2 Seismic Load Effect Including a 2.5 Overstrength Factor

Where specifically required, conditions requiring overstrength factor applications shall be determined in accordance with the following:

1. For use in load combination 5 in Section 2.3.2 or load combinations 5 and 6 in Section 2.4.1, E shall be taken equal to E_m as determined in accordance with Eq. 12.14-7 as follows:

$$E_m = E_{mh} + E_v \tag{12.14-7}$$

2. For use in load combination 7 in Section 2.3.2 or load combination 8 in Section 2.4.1, E shall be taken equal to E_m as determined in accordance with Eq. 12.14-8 as follows:

$$E_m = E_{mh} - E_v \tag{12.14-8}$$

where

E_m = seismic load effect including overstrength factor

E_{mh} = effect of horizontal seismic forces including ~~structural~~ overstrength factor as defined in Section 12.14.3.2.1

E_v = vertical seismic load effect as defined in Section 12.14.3.1.2

12.14.3.2.2 Load Combinations with Overstrength Factor Where the seismic load effect with overstrength factor, E_m, defined in Section 12.14.3.2, is combined with the effects of other loads as set forth in Chapter 2, the following seismic load combinations for structures not subject to flood or atmospheric ice loads shall be used in lieu of the seismic load combinations in Section 2.3.2 or 2.4.1:
(no change in remainder of Section 12.14.3.2.2)

18.6.3 Effective Ductility Demand

The effective ductility demand on the seismic force-resisting system due to the design earthquake, μ_D, and due to the maximum considered earthquake ground motion, μ_M, shall be calculated using Eqs. 18.6-8, 18.6-9, and 18.6-10:

$$\mu_D = \frac{D_{1D}}{D_Y} \geq 1.0 \tag{18.6-8}$$

$$\mu_M = \frac{D_{1M}}{D_Y} \geq 1.0 \tag{18.6-9}$$

$$D_Y = \left(\frac{g}{4\pi^2}\right)\left(\frac{\Omega_0 C_d}{R}\right)\Gamma_1 C_{S1} T_1^2 \tag{18.6-10}$$

where

D_{1D} = fundamental mode design displacement at the center of rigidity of the roof level of the structure in the direction under consideration, Section 18.4.3.2 or 18.5.3.2

D_{1M} = fundamental mode maximum displacement at the center of rigidity of the roof level of the structure in the direction under consideration, Section 18.4.3.5 or 18.5.3.5

D_Y = displacement at the center of rigidity of the roof level of the structure at the effective yield point of the seismic force-resisting system

R = response modification coefficient from Table 12.2-1

C_d = deflection amplification factor from Table 12.2-1

Ω_o = ~~system~~ overstrength factor from Table 12.2-1

(no change in remainder of Section 18.6.3)

18.7.1.3 Combination of Load Effects

The effects on the damping system due to gravity loads and seismic forces shall be combined in accordance with Section 12.4 using the effect of horizontal seismic forces, Q_E, determined in accordance with the analysis. The redundancy factor, ρ, shall be taken equal to 1.0 in all cases, and the seismic load effect with overstrength factor of Section 12.4.3 need not apply to the design of the damping system.

18.7.2.3 Combination of Load Effects

The effects on the damping system and its components due to gravity loads and seismic forces shall be combined in accordance with Section 12.4 using the effect of horizontal seismic forces, Q_E, determined in accordance with Section 18.7.2.5. The redundancy factor, ρ, shall be taken as equal to 1.0 in all cases, and the seismic load effect with overstrength factor of Section 12.4.3 need not apply to the design of the damping system.

Analysis and Commentary

The changes are intended to bring consistency in the use of the term "overstrength factor" throughout ASCE 7-10. Section 12.2.1 contained the only instance of "system overstrength coefficient" in ASCE 7-05. Sections 12.4.3 and 12.14.3.2 contained the only instances of "structural overstrength" in ASCE 7-05. In Sections 12.4.3, 12.4.3.2, 12.14.3.2, 12.14.3.2.2, 18.7.1.3 and 18.7.2.3, "factor" is added after "overstrength" for consistency with Sections 12.4.3.1 and 12.14.3.2.1. The change in Section 12.14.3 is because of an incorrect section reference.

Language Differences for Overstrength Factor

<table>
<tr><th>ASCE 7-05 Language</th><th>ASCE 7-10 Language</th></tr>
<tr><td>structural overstrength</td><td rowspan="5">overstrength factor</td></tr>
<tr><td>overstrength factor</td></tr>
<tr><td>system overstrength coefficient</td></tr>
<tr><td>system overstrength factor</td></tr>
<tr><td>overstrength</td></tr>
</table>

12.2.5.2, 12.3.3.3, 12.3.3.4, 12.3.4.1, 12.10.2.1, 12.13.6.5, 12.13.6.6, 14.1.7, 14.2.3.2.2, 16.1.4, 16.2.4.1

Modification

Seismic Load Effects Including Overstrength Factor

At a Glance

"Load combinations with overstrength factor" is changed consistently to "seismic load effects including overstrength factor." References are revised so that the exception which permits design for the maximum load that can be delivered by an element applies.

2010 Standard

12.2.5.2 Cantilever Column Systems

Cantilever column systems are permitted as indicated in Table 12.2-1 and as follows. The required axial strength of individual cantilever column elements, considering only the load combinations that include seismic load effects, shall not exceed 15 percent of the available axial strength, including slenderness effects.

Foundation and other elements used to provide overturning resistance at the base of cantilever column elements shall be designed ~~have the strength~~ to resist the seismic load effects including ~~load combinations with~~ overstrength factor of Section 12.4.3 ~~12.4.3.2~~.

12.3.3.3 Elements Supporting Discontinuous Walls or Frames

Columns, beams, trusses, or slabs supporting discontinuous walls or frames of structures having horizontal irregularity Type 4 of Table 12.3-1 or vertical irregularity Type 4 of Table 12.3-2 shall be designed ~~have the design strength~~ to resist the seismic load effects including ~~maximum force that can develop in accordance with the load combinations with~~ overstrength factor of Section 12.4.3 ~~12.4.3.2~~. The connections of such discontinuous elements to the supporting members shall be adequate to transmit the forces for which the discontinuous elements were required to be designed.

12.3.3.4 Increase in Forces Due to Irregularities for Seismic Design Categories D through F

For structures assigned to Seismic Design Category D, E, or F and having a horizontal structural irregularity of Type 1a, 1b, 2, 3, or 4 in Table 12.3-1 or a vertical structural irregularity of Type 4 in Table 12.3-2, the design forces determined from Section 12.10.1.1 shall be increased 25 percent for the following elements of the seismic force-resisting system:

1. Connections of diaphragms to vertical elements and to collectors.
2. Collectors and their connections, including connections to vertical elements, of the seismic force-resisting system.

EXCEPTION:

Forces calculated using the seismic load effects including ~~load combinations with~~ overstrength factor of Section 12.4.3 ~~12.4.3.2~~ need not be increased.

12.3.4.1 Conditions Where Value of ρ is 1.0

The value of ρ is permitted to equal 1.0 for the following:

1. Structures assigned to Seismic Design Category B or C.
2. Drift calculation and P-delta effects.
3. Design of nonstructural components.
4. Design of nonbuilding structures that are not similar to buildings.
5. Design of collector elements, splices, and their connections for which the seismic load effects including ~~load combinations with~~ overstrength factor of Section 12.4.3 ~~12.4.3.2~~ are used.
6. Design of members or connections where seismic load effects including ~~load combinations with~~ overstrength factor of Section 12.4.3 ~~12.4.3.2~~ are required for design.
7. Diaphragm loads determined using Eq. 12.10-1.
8. Structures with damping systems designed in accordance with Chapter 18.
9. Design of structural walls for out-of-plane forces, including their anchorage.

12.10.2.1 Collector Elements Requiring Load Combinations with Overstrength Factor for Seismic Design Categories C through F

In structures assigned to Seismic Design Category C, D, E, or F, collector elements (see Fig. 12.10-1) and their connections including connections to vertical elements shall be designed to resist the maximum of the following:

1. Forces calculated using the seismic load effects including ~~load combinations with~~ overstrength factor of Section 12.4.3 ~~12.4.3.2~~ with seismic forces determined by the Equivalent Lateral Force procedure of Section 12.8 or the Modal Response Spectrum Analysis procedure of Section 12.9.
2. Forces calculated using the seismic load effects including ~~load combinations with~~ overstrength factor of Section 12.4.3 ~~12.4.3.2~~ with seismic forces determined by Equation 12.10-1.
3. Forces calculated using the load combinations of Section 12.4.2.3 with seismic forces determined by Equation 12.10-2.

Transfer forces as described in Section 12.10.1.1 shall be considered.

EXCEPTIONS:

(no change in text)

12.13.6.5 Pile Anchorage Requirements

In addition to the requirements of Section 12.13.5.3, anchorage of piles shall comply with this section. Design of anchorage of piles into the pile cap shall consider the combined effect of axial forces due to uplift and bending moments due to fixity to the pile cap. For piles required to resist uplift forces or provide rotational restraint, anchorage into the pile cap shall comply with ~~be capable of developing~~ the following:

1. In the case of uplift, the anchorage shall be capable of developing the least ~~lesser~~ of the nominal tensile strength of the longitudinal reinforcement in a concrete pile, ~~or~~ the nominal tensile strength of a steel pile, and ~~or~~ 1.3 times the pile pullout resistance, or shall be designed to resist the axial tension force resulting from the seismic load effects including ~~load combinations with~~ overstrength factor of Section 12.4.3 ~~12.4.3.2~~ or 12.14.3.2 ~~12.14.3.2.2~~. The pile pullout resistance shall be taken as the ultimate frictional or adhesive force that can be developed between the soil and the pile plus the pile and pile cap weight.
2. In the case of rotational restraint, the anchorage shall be designed to resist the ~~lesser of the~~ axial and shear forces and moments resulting from the seismic load effects including ~~load combinations with~~ overstrength factor of Section 12.4.3 ~~12.4.3.2~~ or 12.14.3.2 ~~12.14.3.2.2~~ or shall be capable of developing ~~development of~~ the full axial, bending, and shear nominal strength of the pile.

12.13.6.6 Splices of Pile Segments

Splices of pile segments shall develop the nominal strength of the pile section.

EXCEPTION: Splices ~~but the splice need not develop the nominal strength of the pile in tension, shear, and bending where it has been~~ designed to resist the axial and shear forces and moments from the seismic load effects including ~~load combinations with~~ overstrength factor of Section 12.4.3 ~~12.4.3.2~~ or 12.14.3.2 ~~12.14.3.2.2~~.

14.1.7 Additional Detailing Requirements for Steel Piles in Seismic Design Categories D through F

In addition to the foundation requirements set forth in Sections 12.1.5 and 12.13, design and detailing of H-piles shall conform to the requirements of AISC 341, and the connection between the pile cap and steel piles or unfilled steel pipe piles in structures assigned to Seismic Design Category D, E, or F shall be designed for a tensile force not less than 10 percent of the pile compression capacity.

EXCEPTION: Connection tensile capacity need not exceed the strength required to resist seismic load effects including ~~load combinations with~~ overstrength factor of Section 12.4.3 ~~12.4.3.2~~ or Section 12.14.3.2 ~~12.14.2.2.2~~. Connections need not be provided where the foundation or supported structure does not rely on the tensile capacity of the piles for stability under the design seismic forces.

14.2.3.2.2 Nonapplicable ACI 318 Sections for Grade Beam and Piles Section 21.12.3.3 of ACI 318 need not apply to ~~where~~ grade beams designed ~~have the required strength~~ to resist the seismic load effects including ~~forces from the load combinations with~~ overstrength factor of Section 12.4.3 ~~12.4.3.2~~ or 12.14.3.2 ~~12.14.3.2.2~~. Section 21.12.4.4(a) of ACI 318 need not apply to concrete piles. Section 21.12.4.4(b) of ACI 318 need not apply to precast, prestressed concrete piles.

16.1.4 Response Parameters

For each ground motion analyzed, the individual response parameters shall be multiplied by the following scalar quantities:

a. Force response parameters shall be multiplied by I_e/R, where I_e is the importance factor determined in accordance with Section 11.5.1 and R is the Response Modification Coefficient selected in accordance with Section 12.2.1.

b. Drift quantities shall be multiplied by C_d/R, where C_d is the deflection amplification factor specified in Table 12.2-1.

For each ground motion i, where i is the designation assigned to each ground motion, the maximum value of the base shear, V_i, member forces, Q_{Ei}, scaled as indicated in the preceding text and story drifts, Δ_i, at each story as defined in Section 12.8.6 shall be determined. Where the maximum scaled base shear predicted by the analysis, V_i, is less than 85 percent of the value of V determined using the minimum value of C_s set forth in Eq. 12.8-5 or when located where S_1 is equal to or greater than $0.6g$, the minimum value of C_s set forth in Eq. 12.8-6, the scaled member forces, Q_{Ei}, shall be additionally multiplied by $\frac{V}{V_i}$ where V is the minimum base shear that has been determined using the minimum value of C_s set forth in Eq. 12.8-5, or when located where S_1 is equal to or greater than $0.6g$, the minimum value of C_s set forth in Eq. 12.8-6. Where the maximum scaled base shear predicted by the analysis, V_i, is less than $0.85C_sW$, where C_s is from Eq. 12.8-6, drifts shall be multiplied by $0.85\,\frac{C_sW}{V_i}$.

If at least seven ground motions are analyzed, the design member forces used in the load combinations of Section 12.4.2.1 and the design story drift used in the evaluation of drift in accordance with Section 12.12.1 are permitted to be taken respectively as the average of the scaled Q_{Ei} and Δ_i values determined from the analyses and scaled as indicated in the preceding text. If fewer than seven ground motions are analyzed, the design member forces and the design story drift shall be taken as the maximum value of the scaled Q_{Ei} and Δ_i values determined from the analyses.

Where this standard requires ~~the~~ consideration of the seismic load effects including ~~load combinations with~~ overstrength factor of Section 12.4.3 ~~12.4.3.2~~, the value of $\Omega_0 Q_E$ need not be taken larger than the maximum of the unscaled value, Q_{Ei}, obtained from the analyses.

16.2.4.1 Member Strength

The adequacy of members to resist the combination of load effects of Section 12.4 need not be evaluated.

EXCEPTION: Where this standard requires ~~the~~ consideration of the seismic load effects including ~~load combinations with~~ overstrength factor of Section 12.4.3 ~~12.4.3.2~~, the maximum value of Q_{Ei} obtained from the suite of analyses shall be taken in place of the quantity $\Omega_0 Q_E$.

Analysis and Commentary

The phrase "load combinations with overstrength factor" is changed to "seismic load effects including overstrength factor" throughout ASCE 7-10 for consistency with the title of Section 12.4.3. In the process of doing this, additional editorial changes that will improve the standard were identified and made.

The references to Section 12.4.3.2 in conjunction with load combinations with overstrength factor were being interpreted by designers as prohibiting use of the exception to Section 12.4.3.1, which permits design to the maximum load that can be developed by an element. Changing the reference section to 12.4.3 clarifies that this exception may be used.

The same exception to Section 12.4.3.1 is also an exception to Section 12.14.3.2.1. Because of this, references to Section 12.14.3.2.2 are changed in all applicable sections of ASCE 7-10 to Section 12.14.3.2.

Olive View Memorial Hospital following the San Fernando earthquake of 1971. The columns supporting the discontinued shear walls are required by ASCE 7-10 to be designed for seismic load effects including overstrength factor.

Courtesy: Earthquake Engineering Research Institute

Part II Chapter 11 Seismic Design Criteria

11.2, 11.8.2, 11.8.3, 21.5.1, 21.5.2, 21.5.3, Chapter 22

Modification and Addition

MCE_G Peak Ground Acceleration, Liquefaction Potential Evaluation

At a Glance

The term, "maximum considered earthquake geometric mean (MCE_G) peak ground acceleration" is defined. The MCE_G peak ground acceleration adjusted for site effects (PGA_M) is used in this standard for evaluation of liquefaction, lateral spreading, seismic settlements, and other soil related issues.

2010 Standard

11.2 DEFINITIONS

The following definitions apply only to the seismic requirements of this standard.

MAXIMUM CONSIDERED EARTHQUAKE (MCE) GROUND MOTION: The most severe earthquake effects considered by this standard more specifically ~~as~~ defined in the following two terms ~~in Section 11.4~~.

MAXIMUM CONSIDERED EARTHQUAKE GEOMETRIC MEAN (MCE_G) PEAK GROUND ACCELERATION: The most severe earthquake effects considered by this standard determined for geometric mean peak ground acceleration and without adjustment for targeted risk. The MCE_G peak ground acceleration adjusted for site effects (PGA_M) is used in this standard for evaluation of liquefaction, lateral spreading, seismic settlements, and other soil related issues. In this standard, general procedures for determining PGA_M are provided in Section 11.8.3; site-specific procedures are provided in Section 21.5.

RISK-TARGETED MAXIMUM CONSIDERED EARTHQUAKE (MCE_R) GROUND MOTION RESPONSE ACCELERATION: The most severe earthquake effects considered by this standard determined for the orientation that results in the largest maximum response to horizontal ground motions and with adjustment for targeted risk. In this standard, general procedures for determining the MCE_R Ground Motion values are provided in Section 11.4.3; site-specific procedures are provided in Sections 21.1 and 21.2.

11.8 GEOLOGIC HAZARDS AND GEOTECHNICAL INVESTIGATION

11.8.1 Site Limitation for Seismic Design Categories E and F

A structure assigned to Seismic Design Category E or F shall not be located where there is a known potential for an active fault to cause rupture of the ground surface at the structure.

11.8.2 Geotechnical Investigation Report Requirements for Seismic Design Categories C through F

A geotechnical investigation report shall be provided for a structure assigned to Seismic Design Category C, D, E, or F in accordance with this section. An investigation shall be conducted and a report shall be submitted that includes an evaluation of the following potential geologic and seismic hazards:

a. Slope instability,

b. Liquefaction,

c. Total and differential settlement, and

d. Surface displacement due to faulting or seismically induced lateral spreading or lateral flow.

The report shall contain recommendations for ~~appropriate~~ foundation designs or other measures to mitigate the effects of the previously mentioned hazards.

EXCEPTION: Where approved ~~deemed appropriate~~ by the authority having jurisdiction, a site-specific geotechnical report is not required where prior evaluations of nearby sites with similar soil conditions provide ~~sufficient~~ direction relative to the proposed construction.

11.8.3 Additional Geotechnical Investigation Report Requirements for Seismic Design Categories D through F

The geotechnical investigation report for a structure assigned to Seismic Design Category D, E, or F shall include all of the following, as applicable:

1. The determination of dynamic seismic lateral earth pressures on basement and retaining walls due to design earthquake ground motions.
2. The potential for liquefaction and soil strength loss evaluated for site peak ground accelerations~~,~~ earthquake magnitudes~~,~~ and source characteristics consistent with the MCE_G peak ground acceleration ~~design earthquake ground motions~~. Peak ground acceleration shall ~~is permitted to~~ be determined based on either (1) a site-specific study taking into account soil amplification effects as specified in Section 11.4.7, or, ~~in the absence of such a study,~~ (2) the peak ground accelerations ~~shall be assumed equal to~~ ~~$S_s/2.5$~~, PGA_M, from Eq. 11.8-1.

$$PGA_M = F_{PGA} PGA \quad \text{(Eq. 11.8-1)}$$

where

PGA_M = MCE_G peak ground acceleration adjusted for Site Class effects.

PGA = Mapped MCE_G peak ground acceleration shown in Figs. 22-7 through 22-11.

F_{PGA} = Site coefficient from Table 11.8-1.

3. Assessment of potential consequences of liquefaction and soil strength loss, including, but not limited to, estimation of total and differential settlement, lateral soil movement, lateral soil loads on foundations, reduction in foundation soil-bearing capacity and lateral soil reaction, soil downdrag and reduction in axial and lateral soil reaction for pile foundations, increases in soil lateral pressures on retaining walls, and flotation of buried structures.
4. Discussion of mitigation measures such as, but not limited to, ~~ground stabilization~~ selection of appropriate foundation type and depths, selection of appropriate structural systems to accommodate anticipated displacements and forces, ground stabilization, or any combination of these measures and how they shall be considered in the design of the structure.

Table 11.8-1 Site Coefficient F_{PGA}

Site Class	Mapped Maximum Considered Geometric Mean (MCE_G) Peak Ground Acceleration, PGA				
	PGA ≤ 0.1	PGA = 0.2	PGA = 0.3	PGA = 0.4	PGA ≥ 0.5
A	0.8	0.8	0.8	0.8	0.8
B	1.0	1.0	1.0	1.0	1.0
C	1.2	1.2	1.1	1.0	1.0
D	1.6	1.4	1.2	1.1	1.0
E	2.5	1.7	1.2	0.9	0.9
F	See Section 11.4.7				

Note: Use straight-line interpolation for intermediate values of PGA.

21.5 MAXIMUM CONSIDERED EARTHQUAKE GEOMETRIC MEAN (MCE_G) PEAK GROUND ACCELERATION

21.5.1 Probabilistic MCE_G Peak Ground Acceleration

The probabilistic geometric mean peak ground acceleration shall be taken as the geometric mean peak ground acceleration with a 2 percent probability of exceedance within a 50-year period.

21.5.2 Deterministic MCE_G Peak Ground Acceleration

The deterministic geometric mean peak ground acceleration shall be calculated as the largest 84th-percentile geometric mean peak ground acceleration for characteristic earthquakes on all known active faults within the site region. The deterministic geometric mean peak ground acceleration shall not be taken as lower than 0.5 F_{PGA}, where F_{PGA} is determined using Table 11.8-1 with the value of PGA taken as 0.5g.

21.5.3 Site-Specific MCE_G Peak Ground Acceleration

The site-specific MCE_G peak ground acceleration, PGA_M, shall be taken as the lesser of the probabilistic geometric mean peak ground acceleration of Section 21.5.1 and the deterministic geometric mean peak ground acceleration of Section 21.5.2. The site-specific MCE_G peak ground acceleration shall not be taken less than 80 percent of PGA_M determined from Eq. 11.8-1.

Note: For brevity, Chapter 22 is not included here. For the text of Chapter 22, see ASCE 7-10 or page 41. It is in this chapter that the MCE_G peak ground acceleration maps are found (see Figures 22-7 through 22-11).

Analysis and Commentary

Section 11.8.3 Item 2 is modified to require that evaluations of liquefaction potential be made for maximum considered earthquake (MCE) ground motions rather than design earthquake ground motions to assure that the potential occurrence and effects of liquefaction during the MCE are considered in geotechnical and structural design. This change is consistent with the adoption of a risk-based target for collapse prevention as a performance goal as well as with other evaluations for the MCE required by the NEHRP *Provisions* when judged necessary to meet the collapse-prevention performance goal during MCE loading.

The provision also requires that liquefaction potential evaluations be conducted using mapped peak ground acceleration values (maps provided in Figures 22-7 through 22-11) adjusted for site effects, rather than using the current approximation for peak ground acceleration of short-period spectral acceleration multiplied by a factor of 0.4. The new maps provide substantially more accurate values for PGA, since they are based on PGA attenuation relationships. PGA is modified for site class effects using Eq. 11.8-1, where the site coefficient F_{PGA} is obtained from Table 11.8-1. Values of F_{PGA} in the table are identical to those of F_a in Table 11.4-1 but are a function of PGA rather than S_S. Because PGA is a short-period parameter (equal to zero-period spectral acceleration), it is appropriate and consistent with current practice to use the same site coefficients for PGA and S_S. It is also consistent with the original development of F_a as a function of PGA.

The mapped peak ground accelerations in Figure 22-7 through 22-11 are geomean values and not risk-targeted values. Thus these are designated as MCE_G peak ground accelerations, unlike the spectral accelerations in Figures 22-1 through 22-6, which represent risk-targeted MCE or MCE_R ground motion.

The newly added Sections 21.5.1 (Probabilistic MCE_G Peak Ground Acceleration), 21.5.2 (Deterministic MCE_G Peak Ground Acceleration), and 21.5.3 (Site-Specific MCE_G Peak Ground Acceleration) parallel Sections 21.2.1 (Probabilistic (MCE_R) Ground Motions), 21.2.2 (Deterministic (MCE_R) Ground Motions), and 21.2.3 (Site-Specific MCE_R), respectively. In Section 21.5.2, 0.5 F_{PGA} is the deterministic lower limit on PGA, where 0.5*g* is the bedrock PGA and F_{PGA} is the site coefficient. A bedrock PGA of 0.6*g* would have been the exact equivalent of the lower-bound limits of 1.5*g* and 0.6*g* on S_S and S_1, respectively in Section 21.2.2. There was some objection to the 0.6*g* lower limit as putting a constraint on liquefaction analysis, where there had previously been no limit. Some felt that the lower bounds on S_S and S_1 had their origin in structural behavior and should not apply to liquefaction. The 0.5*g* (rather than 0.6*g*) was considered more appropriate as the lower limit on bedrock acceleration, based on discussions within the Seismic Subcommittee of ASCE 7.

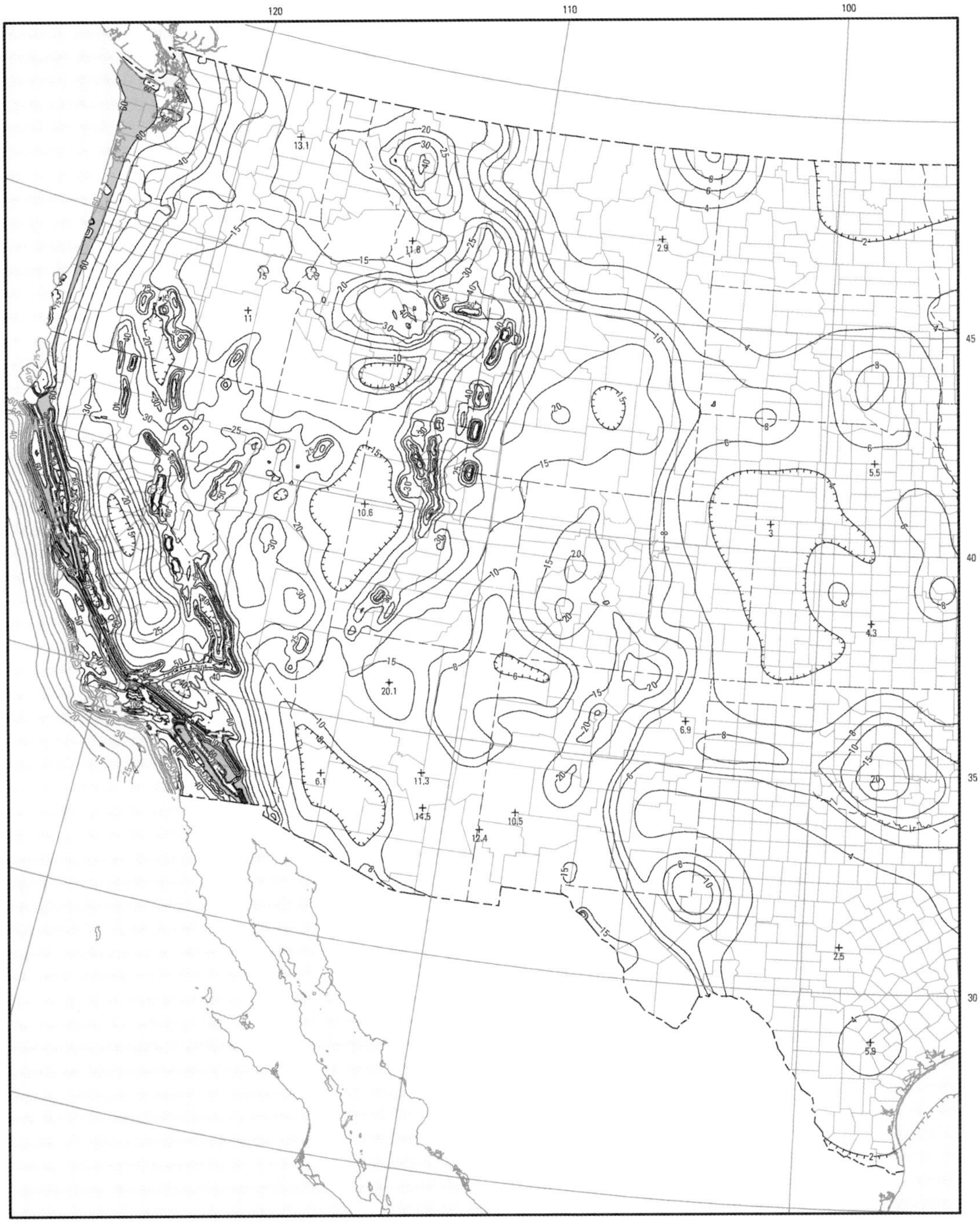

Figure 22-7 (Partial) Maximum Considered Earthquake Geometric Mean (MCE_G) PGA, %*g*, Site Class B for the Western United States.

1.4.2, 1.4.3, 1.4.4, 1.4.5, 11.7.1–11.7.5 Modification and Relocation

Design Requirements for Seismic Design Category A

At a Glance

Much of the contents of Section 11.7 has been transferred to Section 1.4, General Structural Integrity.

2010 Standard

11.7 DESIGN REQUIREMENTS FOR SEISMIC DESIGN CATEGORY A

~~**11.7.1 Applicability of Seismic Requirements for Seismic Design Category A Structures.**~~ Buildings and other s~~S~~tructures ~~and their components~~ ~~Structures~~ assigned to Seismic Design Category A need only comply with the requirements of Section ~~11.7~~ 1.4. Nonstructural components in SDC A are exempt from seismic design requirements. In addition, tanks assigned to Risk Category IV, shall satisfy the freeboard requirement in Section 15.7.6.1.2. ~~The effects on the structure and its components due to the forces prescribed in this section shall be taken as E and combined with the effects of other loads in accordance with the load combinations of Section 2.3 or 2.4. For structures with damping systems, see Section 18.2.1.~~

Delete Sections 11.7.2 through 11.7.5 (text not shown here)

1.4.2 Load Path Connections

All parts of the structure between separation joints shall be interconnected to form a continuous path to the lateral force-resisting system, and the connections shall be capable of transmitting the lateral forces induced by the parts being connected. Any smaller portion of the structure shall be tied to the remainder of the structure with elements having strength to resist a force of not less than 5% of the portion's weight.

1.4.3 Lateral Forces

Each structure shall be analyzed for the effects of static lateral forces applied independently in each of two orthogonal directions. In each direction, the static lateral forces at all levels shall be applied simultaneously. For purposes of analysis, the force at each level shall be determined using Eq. 1.4-1 as follows:

$$F_x = 0.01W_x \qquad (1.4\text{-}1)$$

where

F_x = the design lateral force applied at story x and

W_x = the portion of the total dead load of the structure, D, located or assigned to level x.

Structures explicitly designed for stability, including second-order effects, shall be deemed to comply with the requirements of this section.

1.4.4 Connection to Supports

A positive connection for resisting a horizontal force acting parallel to the member shall be provided for each beam, girder, or truss either directly to its supporting elements, or to slabs designed to act as diaphragms. Where the connection is through a diaphragm, the member's supporting element shall also be connected to the diaphragm. The connection shall have the strength to resist a force of 5 percent of the unfactored dead plus live load reaction imposed by the supported member on the supporting member.

1.4.5 Anchorage of Structural Walls

Walls that provide vertical load bearing or lateral shear resistance for a portion of the structure shall be anchored to the roof and all floors and members that provide lateral support for the wall or that are supported by the wall. The anchorage shall provide a direct connection between the walls and the roof or floor construction. The connections shall be capable of resisting a strength level horizontal force perpendicular to the plane of the wall equal to 0.2 times the weight of the wall tributary to the connection, but not less than 5 psf (0.24 kN/m^2).

Analysis and Commentary

Much of the contents of Section 11.7, Design Requirements for Seismic Design Category A, has been transferred in modified form to Section 1.4, General Structural Integrity. The latter was considered to be a more logical location. The modifications are beyond the scope here. The seismic modifications are discussed as part of other changes.

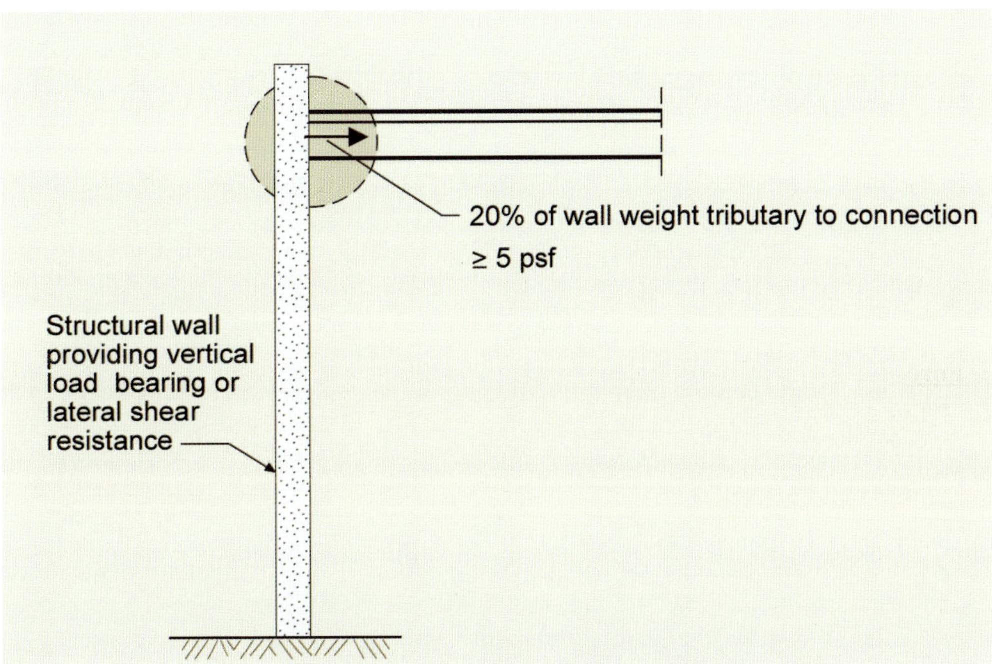

Illustration of Section 1.4.5 Anchorage of Structural Walls
Courtesy: International Code Council

11.1.3 Clarification

What Qualifies as a Nonbuilding Structure?

At a Glance

Language is added to clarify types of buildings that may be classified as nonbuilding structures in certain situations.

2010 Standard

11.1.3 Applicability

Structures and their nonstructural components shall be designed and constructed in accordance with the requirement of the following sections based on the type of structure or component:

a. Buildings: Chapter 12

b. Nonbuilding Structures: Chapter 15

c. Nonstructural Components: Chapter 13

d. Seismically Isolated Structures: Chapter 17

e. Structures with Damping Systems: Chapter 18

Buildings whose purpose is to enclose equipment or machinery and whose occupants are engaged in maintenance or monitoring of that equipment, machinery or their associated processes shall be permitted to be classified as nonbuilding structures designed and detailed in accordance with Section 15.5 of this standard.

Analysis and Commentary

This change was initiated at an AISC- sponsored Industrial Building and Seismic Design Summit Meeting held in January 2009. The purpose of the meeting was to find safe and suitable solutions to address the problems industrial buildings faced in meeting the system and height limitations set forth in Table 12.2-1.

This particular change permits industrial buildings to be designed by the nonbuilding structures requirements in Chapter 15. Contrary to its appearance, the last paragraph of Section 11.1.3 is not a substantive change to previous requirements, but rather a clarification as to which structures can be classified as nonbuilding structures.

Language has been added to Section 11.1.3 of the ASCE 7-10 Commentary, which states that examples of such structures include, but are not limited to, boiler buildings, aircraft hangars, steel mills, aluminum smelting facilities, and other automated manufacturing facilities. The occupancy restrictions for such facilities should be uniquely reviewed in each case.

These types of buildings are permitted to be classified as nonbuilding structures.

11.2, 11.3, 11.4.1, 11.4.3, 11.4.5, 11.4.6, 21.2.1, 21.2.1.1, 21.2.1.2, 21.2.2, 21.2.3, Chapter 22, Figures 22-1 through 22-6

Modification

New Seismic Ground Motion Maps and Related Technical Changes

At a Glance

Seismic ground motion maps have been changed to include: (1) new information on earthquake source and ground motion prediction equations, (2) risk-targeted ground motions, (3) maximum direction ground motions, and (4) near-source 84th percentile ground motions.

2010 Standard

11.2 DEFINTIONS

DESIGN EARTHQUAKE: The earthquake effects that are two-thirds of the corresponding Maximum Considered Earthquake (MCE_R) effects.

DESIGN EARTHQUAKE GROUND MOTION: The earthquake ground motions that are two-thirds of the corresponding MCE_R ground motions.

MAXIMUM CONSIDERED EARTHQUAKE (MCE) GROUND MOTION: The most severe earthquake effects considered by this standard more specifically ~~as~~ defined in the following two terms ~~in Section 11.4~~.

MAXIMUM CONSIDERED EARTHQUAKE GEOMETRIC MEAN (MCE_G) PEAK GROUND ACCELERATION: The most severe earthquake effects considered by this standard determined for geometric mean peak ground acceleration and without adjustment for targeted risk. The MCE_G peak ground acceleration adjusted for site effects (PGA_M) is used in this standard for evaluation of liquefaction, lateral spreading, seismic settlements, and other soil related issues. In this standard, general procedures for determining PGA_M are provided in Section 11.8.3; site-specific procedures are provided in Section 21.5.

RISK-TARGETED MAXIMUM CONSIDERED EARTHQUAKE (MCE_R) GROUND MOTION RESPONSE ACCELERATION: The most severe earthquake effects considered by this standard determined for the orientation that results in the largest maximum response to horizontal ground motions and with adjustment for targeted risk. In this standard, general procedures for determining the MCE_R Ground Motion values are provided in Section11.4.3; site-specific procedures are provided in Sections 21.1 and 21.2.

11.3 SYMBOLS

The unit dimensions used with the items covered by the symbols shall be consistent throughout except where specifically noted. Symbols presented in this section apply only to the seismic requirements in this standard as indicated.

C_R = site-specific risk coefficient at any period; see Section 21.2.1.1

C_{RS} = mapped value of the risk coefficient at short periods as given by Fig. 22-17

C_{R1} = mapped value of the risk coefficient at a period of 1 s as given by Fig. 22-18

S_S = mapped MCE$_R$, 5 percent damped, spectral response acceleration parameter at short periods as defined in Section 11.4.1

S_1 = mapped MCE$_R$, 5 percent damped, spectral response acceleration parameter at a period of 1 s as defined in Section 11.4.1

S_{aM} = the site-specific MCE$_R$ spectral response acceleration parameter at any period

S_{MS} = the MCE$_R$, 5 percent damped, spectral response acceleration parameter at short periods adjusted for site class effects as defined in Section 11.4.3

S_{M1} = the MCE$_R$, 5 percent damped, spectral response acceleration parameter at a period of 1 s adjusted for site class effects as defined in Section 11.4.3

11.4.1 Mapped Acceleration Parameters.

The parameters S_S and S_1 shall be determined from the 0.2 and 1 s spectral response accelerations shown on Figs. 22-1, 22-3, 22-5, and 22-6 for S_s and Figs. 22-2, 22-4, 22-5, and 22-6 for S_1 ~~Figures 22-1 through 22-14, respectively~~. Where S_1 is less than or equal to 0.04 and S_S is less than or equal to 0.15, the structure is permitted to be assigned to Seismic Design Category A and is only required to comply with Section 11.7.

User Note: Electronic values of mapped acceleration parameters, and other seismic design parameters, are provided at the USGS Web site at http://earthquake.usgs.gov/designmaps, or through the SEI Web site at http://content.seinstitute.org.

11.4.3 Site Coefficients and Risk-Targeted ~~Adjusted~~ Maximum Considered Earthquake (MCE$_R$) Spectral Response Acceleration Parameters

The MCE$_R$ spectral response acceleration parameter for short periods (S_{MS}) and at 1 s (S_{M1}), adjusted for Site Class effects, shall be determined by Eqs. 11.4-1 and 11.4-2, respectively.

$$S_{MS} = F_a S_S \quad (11.4\text{-}1)$$

$$S_{M1} = F_v S_1 \quad (11.4\text{-}2)$$

where

S_S = the mapped MCE$_R$ spectral response acceleration parameter at short periods as determined in accordance with Section 11.4.1, and

S_1 = the mapped MCE$_R$ spectral response acceleration parameter at a period of 1 s as determined in accordance with Section 11.4.1

where site coefficients F_a and F_v are defined in Tables 11.4-1 and 11.4-2, respectively. Where the simplified design procedure of Section 12.14 is used, the value of F_a shall be determined in accordance with Section 12.14.8.1, and the values for F_v, S_{MS}, and S_{M1} need not be determined.

11.4.5 Design Response Spectrum.

Where a design response spectrum is required by this standard and site-specific ground motion procedures are not used, the design response spectrum curve shall be developed as indicated in Fig. 11.4-1 and as follows:

1. For periods less than T_0, the design spectral response acceleration, S_a, shall be taken as given by Eq. 11.4-5:

$$S_a = S_{DS}\left(0.4 + 0.6\frac{T}{T_0}\right) \qquad (11.4\text{-}5)$$

2. For periods greater than or equal to T_0 and less than or equal to T_S, the design spectral response acceleration, S_a, shall be taken equal to S_{DS}.
3. For periods greater than T_S, and less than or equal to T_L, the design spectral response acceleration, S_a, shall be taken as given by Eq. 11.4-6:

$$S_a = \frac{S_{D1}}{T} \qquad (11.4\text{-}6)$$

4. For periods greater than T_L, S_a shall be taken as given by Eq. 11.4-7:

$$S_a = \frac{S_{D1}T_L}{T^2} \qquad (11.4\text{-}7)$$

where

S_{DS} = the design spectral response acceleration parameter at short periods

S_{D1} = the design spectral response acceleration parameter at 1-s period

T = the fundamental period of the structure, s

T_0 = $0.2\dfrac{S_{D1}}{S_{DS}}$

T_S = $\dfrac{S_{D1}}{S_{DS}}$ and

T_L = long-period transition period (s) shown in Figs 22-12 through 22-16 .~~22-15 (Conterminous United States), Figure 22--16 (Region 1), Figure 22--17 (Alaska), Figure 22--18 (Hawaii), Figure 22--19 (Puerto Rico, Culebra, Vieques, St. Thomas, St. John, and St. Croix), and Figure 22--20 (Guam and Tutuila).~~

11.4.6 Risk-Targeted Maximum Considered (MCE$_R$) Response Spectrum

Where an MCE$_R$ response spectrum is required, it shall be determined by multiplying the design response spectrum by 1.5.

21.2.1 Probabilistic (MCE$_R$) Ground Motions ~~MCE~~

The probabilistic ~~MCE~~ spectral response accelerations shall be taken as the spectral response accelerations in the direction of maximum horizontal response represented by a 5 percent damped acceleration response spectrum that is expected to achieve a 1 percent probability of collapse ~~having a 2 percent probability of exceedance~~ within a 50-year period. For the purpose of this standard, ordinates of the probabilistic ground motion response spectrum shall be determined by either Method 1 of Section 21.2.1.1 or Method 2 of Section 21.2.1.2.

21.2.1.1 Method 1

At each spectral response period for which the acceleration is computed, ordinates of the probabilistic ground motion response spectrum shall be determined as the product of the risk coefficient, C_R, and the spectral response acceleration from a 5 percent damped acceleration response spectrum having a 2 percent probability of exceedance within a 50-year period. The value of the risk coefficient, C_R, shall be determined using values of C_{RS} and C_{R1} from Figs. 22-17 and 22-18, respectively. At spectral response periods less than or equal to 0.2 s, C_R shall be taken as equal to C_{RS}. At spectral response periods greater than or equal to 1.0 s, C_R shall be taken as equal to C_{R1}. At response spectral periods greater than 0.2 s and less than 1.0 s, C_R shall be based on linear interpolation of C_{RS} and C_{R1}.

21.2.1.2 Method 2

At each spectral response period for which the acceleration is computed, ordinates of the probabilistic ground motion response spectrum shall be determined from iterative integration of a site-specific hazard curve with a lognormal probability density function representing the collapse fragility (i.e., probability of collapse as a function of spectral response acceleration). The ordinate of the probabilistic ground motion response spectrum at each period shall achieve a 1 percent probability of collapse within a 50-year period for a collapse fragility having (i) a 10 percent probability of collapse at said ordinate of the probabilistic ground motion response spectrum and (ii) a logarithmic standard deviation value of 0.6.

21.2.2 Deterministic (MCE_R) Ground Motions ~~MCE~~.

The deterministic spectral ~~MCE~~ response acceleration at each period shall be calculated as an 84th-percentile ~~150 percent of the largest median~~ 5 percent damped spectral response acceleration in the direction of maximum horizontal response computed at that period. The largest such acceleration calculated for the characteristic earthquakes on all known active faults within the region shall be used. For the purposes of this standard, the ordinates of the deterministic ~~MCE~~ ground motion response spectrum shall not be taken as lower than the corresponding ordinates of the response spectrum determined in accordance with Fig. 21.2-1, where F_a and F_v are determined using Tables 11.4-1 and 11.4-2, respectively, with the value of S_S taken as 1.5 and the value of S_1 taken as 0.6.

21.2.3 Site-Specific MCE_R.

The site-specific MCE_R spectral response acceleration at any period, S_{aM}, shall be taken as the lesser of the spectral response accelerations from the probabilistic ground motions ~~MCE~~ of Section 21.2.1 and the deterministic ground motions ~~MCE~~ of Section 21.2.2.

For clarity, Chapter 22 is not shown in strikeout and underline format. The following is ASCE 7-10 language. The terms "risk-adjusted" and "mapped MCE_R" as used in ASCE 7-10 are corrected to "risk-targeted" and "risk-targeted MCE_R", respectively, per ASCE 7-10 Errata. For brevity, the new MCE_G peak ground acceleration maps are not reprinted here.
See ASCE 7-10 Chapter 22 or the USGS website.

Chapter 22
SEISMIC GROUND MOTION LONG-PERIOD TRANSITION AND RISK COEFFICIENT MAPS

Contained in this chapter are Figs. 22-1 through 22-6, which provide the risk-targeted maximum considered earthquake (MCE_R) ground motion parameters S_S and S_1; Figs. 22-17 and 22-18 which provide the risk coefficients C_{RS} and C_{R1}, and Figs. 22-12 through 22-15, which provide the long-period transition periods T_L for use in applying the seismic provisions of this standard. S_S is the risk-targeted MCE_R, 5 percent damped, spectral response acceleration parameter at short periods as defined in Section 11.4.1. S_1 is the risk-targeted MCE_R ground motion, 5 percent damped, spectral response acceleration parameter at a period of 1 s as defined in Section 11.4.1. C_{RS} is the mapped risk coefficient at short periods used in Section 21.2.1.1. C_{R1} is the mapped risk coefficient at a period of 1 s used in Section 21.2.1.1. T_L is the mapped long-period transition period used in Section 11.4.5.

These maps were prepared by the United States Geological Survey (USGS) in collaboration with the Building Seismic Safety Council (BSSC) Seismic Design Procedures Reassessment Group and the American Society of Civil Engineers (ASCE) 7 Seismic Subcommittee and have been updated for the 2010 edition of this standard.

Maps of the MCE_R ground motion parameters, S_S and S_1, for Guam and American Samoa are not provided because parameters have not yet been developed for those islands. Therefore, as in the 2005 edition of this standard, the parameters S_S and S_1 shall be, respectively, 1.5 and 0.6 for Guam and 1.0 and 0.4 for American Samoa. Maps of the mapped risk coefficients, C_{RS} and C_{R1}, are also not provided.

Also contained in this chapter are Figs. 22-7 through 22-11, which provide the maximum considered earthquake geometric mean (MCE_G) peak ground acceleration as a percentage of g for Site Class B.

Analysis and Commentary

Incorporated through this change are new seismic hazard data developed by the United States Geological Survey (USGS) National Seismic Hazard Mapping Project and related technical changes developed by the Seismic Design Procedures Reassessment Group (SDPRG) of the Provisions Update Committee (PUC) of the Building Seismic Safety Council (BSSC).

The USGS and the SDPRG worked together to update the seismic design maps and procedures for the 2009 NEHRP *Provisions*. Four significant changes are made. 1) USGS has updated some source zone models, have used Next Generation Attenuation (NGA) relationships, to the exclusion of the old attenuation relationships, in the western United States, and have used new attenuation relationships in addition to the old relationships in the central and eastern United States. The new relationships apparently show that eastern earthquakes are much more like western earthquakes than was thought earlier, with ground motion intensity dropping off more steeply with distance from the source than indicated by earlier attenuation curves. As a result of all of this, ground motion– particularly long-period ground motion– has decreased significantly (by 50% or more) in many parts of the United States. 2) Uniform-hazard ground motion has now been replaced by risk-targeted ground motion. This switch from a 2% in 50-year hazard level to a 1% in 50-year collapse risk target has resulted in up to 30% decreases in ground motion in high-hazard areas of the central and eastern United States and in coastal Oregon. 3) A switch has been made from "geo-mean" ground motions (square root of the product of ground motions in any two orthogonal directions) to maximum direction ground motions. This has resulted in increases in short-period ground motion by a factor of 1.1 and in long-period ground motion by a factor of 1.3. 4) Finally, deterministic ground motions have been changed from 150% of median ground motions to 84th percentile ground motions, which are 180% of median ground motions. The net result of these four major changes has been that short-period ground motions have gone down rather substantially in the central and eastern United States; elsewhere, status quo has by and large been maintained.

Two additional changes were made by the BSSC PUC for the 2009 NEHRP *Provisions*: (1) renaming Maximum Considered Earthquake (MCE) ground motions as Risk-Targeted Earthquake (RTE) ground motions and (2) using formulas that added transparency to the development of RTE ground motions. The first change is incorporated in a modified form. The abbreviation RTE is not used in ASCE 7-10. The term " Risk-Targeted Maximum Considered Earthquake Ground Motion Response Acceleration" is defined; however, the notation used to represent this term is MCE_R. The second change is not incorporated into ASCE 7-10. The equations added for transparency and the related maps are consolidated or removed, so that users would continue to use the same equations in Section 11.4 as those of ASCE 7-05.

STEPS INVOLVED IN THE CREATION OF ASCE 7-10 SEISMIC MAPS

Step 1 - Adjust ASCE 7-05 map which is based on uniform-hazard ground motions (Site Class B) for target risk of collapse

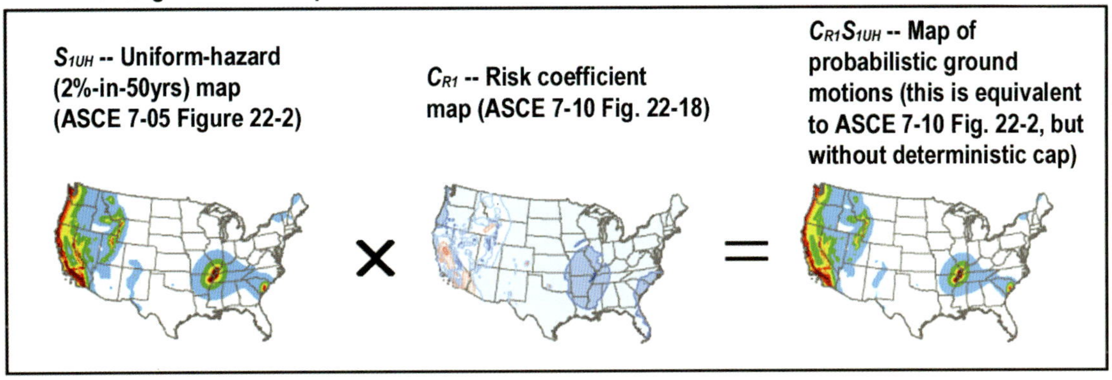

Step 2 - Take minimum of probabilistic and deterministic ground motions (Site Class B) and create map

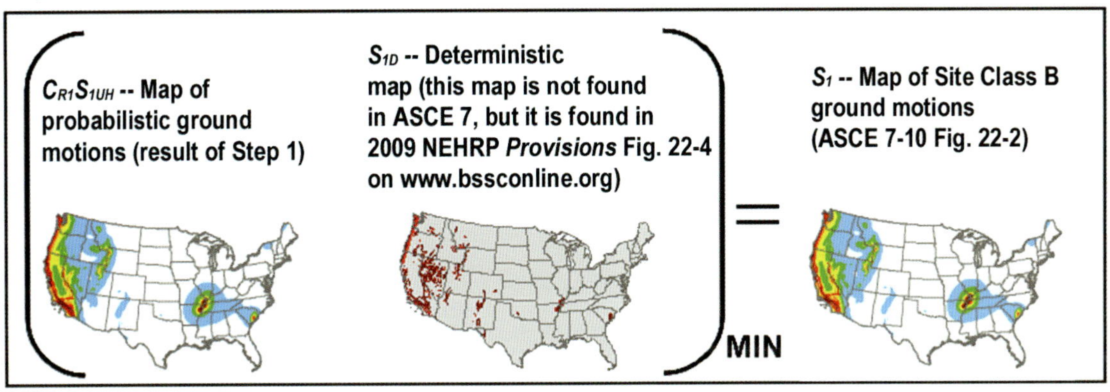

Step 3 - Adjust Site Class B ground motions for site conditions (e.g. Site Class D)

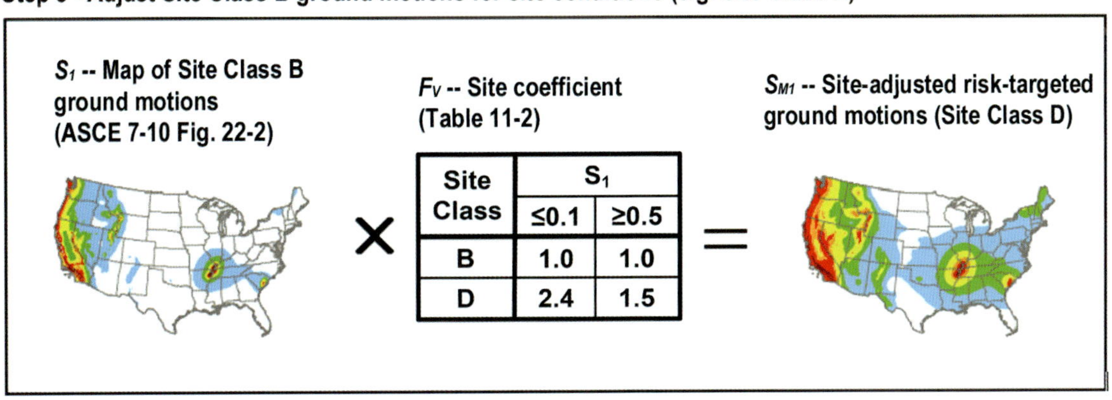

Site Class	S_1	
	≤0.1	≥0.5
B	1.0	1.0
D	2.4	1.5

Part III
Chapter 12 Seismic Design Requirements for Building Structures

Table 12.2-1 Modification

Design Coefficients and Factors for Seismic Force-Resisting Systems of Steel and Wood other than 1) Cold-Formed Steel – Special Bolted Moment Frames and 2)Cantilevered Column Systems (both discussed separately in subsequent items)

At a Glance

Several modifications have been made in the entries for seismic force-resisting systems of steel and wood in Table 12.2-1. The changes are sometimes in the system descriptions, sometimes in the detailing requirements section references, and other times in the design coefficients and factors.

2010 Standard

Table 12.2-1: Design Coefficients and Factors for Seismic Force-Resisting Systems

Seismic Force-Resisting System	ASCE 7 Section Where Detailing Requirements are Specified	Response Modification Coefficient, R^a	Overstrength Factor, Ω_0^g	Deflection Amplification Factor, C_d^b	Structural System Limitations including Structural Height, h_n (ft) Limits[c]				
					Seismic Design Category				
					B	C	D[d]	E[d]	F[e]
(only entries with revisions are shown)									
A. BEARING WALL SYSTEMS									
~~13. Light-framed walls sheathed with wood structural panels rated for shear resistance or steel sheets~~	~~14.1, 14.1.4.2, and 14.5~~	~~6 ½~~	~~3~~	~~4~~	~~NL~~	~~NL~~	~~65~~	~~65~~	~~65~~
15. Light-frame (wood) walls sheathed with wood structural panels rated for shear resistance	14.5	6 ½	3	4	NL	NL	65	65	65
16. Light-frame (cold-formed steel) walls sheathed with wood structural panels rated for shear resistance or steel sheets	14.1	6 ½	3	4	NL	NL	65	65	65
17 ~~14~~. Light-frame~~d~~ walls with shear panels of all other materials	14.1~~,~~ ~~14.1.4.2,~~ and 14.5	2	2 ½	2	NL	NL	35	NP	NP
18. ~~15.~~ Light-frame~~d~~ (cold-formed steel) wall systems using flat strap bracing	14.1, ~~14.1.4.2, and 14.5~~	4	2	3 ½	NL	NL	65	65	65
B. BUILDING FRAME SYSTEMS									
1. Steel eccentrically braced frames~~, moment resisting connections at columns away from links.~~	14.1	8	2	4	NL	NL	160	160	100

Seismic Force-Resisting System	ASCE 7 Section Where Detailing Requirements are Specified	Response Modification Coefficient, R^a	Overstrength Factor, $\Omega_0{}^g$	Deflection Amplification Factor, $C_d{}^b$	Structural System Limitations including Structural Height, h_n (ft) Limits[c]				
					Seismic Design Category				
					B	C	D[d]	E[d]	F[e]
~~2. Steel eccentrically braced frames, non-moment resisting, connections at columns away from links.~~	~~14.1~~	~~7~~	~~2~~	4	~~NL~~	~~NL~~	~~160~~	~~160~~	~~100~~
2.~~3.~~ Steel special ~~Steel~~ concentrically braced frames	14.1	6	2	5	NL	NL	160	160	100
3.~~4.~~ Steel ordinary ~~Steel~~-concentrically braced frames	14.1	3 ¼	2	3 ¼	NL	NL	35[j]	35[j]	NP[j]
22 ~~23~~. Light-frame (wood) walls sheathed with wood structural panels rated for shear resistance ~~or steel sheets~~	~~14.1, 14.1.4.2, and~~ 14.5	7	2 ½	4 ½	NL	NL	65	65	65
23. Light-frame (cold-formed steel) walls sheathed with wood structural panels rated for shear resistance or steel sheets	14.1	7	2 ½	4 ½	NL	NL	65	65	65
24. Light-frame~~d~~ walls with shear panels of all other materials	14.1, ~~14.1.4.2,~~ and 14.5	2 ½	2 ½	2 ½	NL	NL	35	NP	NP
~~25. Buckling restrained braced frames, non-momnt resisting beam column connections~~	~~14.1~~	~~7~~	~~2~~	~~5 ½~~	~~NL~~	~~NL~~	~~160~~	~~160~~	~~100~~
25. Steel buckling-restrained braced frames~~, moment resisting beam column connections~~	14.1	8	2 ½	5	NL	NL	160	160	100
26. Steel special ~~steel~~ plate shear walls	14.1	7	2	6	NL	NL	160	160	100
C. MOMENT-RESISTING FRAME SYSTEMS									
1. Steel special ~~steel~~ moment frames	14.1 and 12.2.5.5	8	3	5 ½	NL	NL	NL	NL	NL
2. Steel special ~~steel~~ truss moment frames	14.1	7	3	5 ½	NL	NL	160	100	NP
3. Steel intermediate ~~steel~~ moment frames	~~12.2.5.6,~~ 12.2.5.7~~,~~ ~~12.2.5.8,~~ ~~12.2.5.9,~~ and 14.1	4 ½	3	4	NL	NL	35[h,~~i~~]	NP[h]	NP[h,i]
4. Steel ordinary ~~steel~~ moment frames	12.2.5.6~~,~~ ~~12.2.5.7,~~ ~~12.2.5.8,~~ and 14.1	3 ½	3	3	NL	NL	NP[~~h~~ i]	NP[~~h~~ i]	NP[i]

Table 12.2-1 | Design Coefficients and Factors for Seismic Force-Resisting Systems of Steel and Wood

Seismic Force-Resisting System	ASCE 7 Section Where Detailing Requirements are Specified	Response Modification Coefficient, R^a	Overstrength Factor, $\Omega_0{}^g$	Deflection Amplification Factor, $C_d{}^b$	Structural System Limitations including Structural Height, h_n (ft) Limits[c]				
					Seismic Design Category				
					B	C	D[d]	E[d]	F[e]
8. Steel and concrete composite special ~~composite steel and concrete~~ moment frames	12.2.5.5 and 14.3	8	3	5 ½	NL	NL	NL	NL	NL
9. Steel and concrete composite intermediate ~~composite~~ moment frames	14.3	5	3	4 ½	NL	NL	NP	NP	NP
10. Steel and concrete composite partially restrained moment frames	14.3	6	3	5 ½	160	160	100	NP	NP
11. Steel and concrete composite ordinary ~~composite~~ moment frames	14.3	3	3	2 ½	NL	NP	NP	NP	NP
D. DUAL SYSTEMS WITH SPECIAL MOMENT FRAMES CAPABLE OF RESISTING AT LEAST 25% OF PRESCRIBED SEISMIC FORCES	12.2.5.1								
2. Steel special concentrically braced frames	14.1	7	2 ½	5 ½	NL	NL	NL	NL	NL
5. Steel and concrete composite ~~steel and concrete~~ eccentrically braced frames	14.3	8	2 ½	4	NL	NL	NL	NL	NL
6. ~~Composite~~ Steel and concrete composite special concentrically braced frames	14.3	6	2 ½	5	NL	NL	NL	NL	NL
7. Steel and concrete composite ~~steel~~ plate shear walls	14.3	7 ½	2 ½	6	NL	NL	NL	NL	NL
8. Steel and concrete composite special ~~composite reinforced concrete~~ shear walls ~~with steel elements~~	14.3	7	2 ½	6	NL	NL	NL	NL	NL
9. Steel and concrete composite ordinary ~~composite reinforced concrete~~ shear walls ~~with steel elements~~	14.3	6	2 ½	5	NL	NL	NP	NP	NP
12. Steel buckling-restrained braced frames	14.1	8	2 ½	5	NL	NL	NL	NL	NL
13. Steel special ~~steel~~ plate shear walls	14.1	8	2 ½	6 ½	NL	NL	NL	NL	NL

Seismic Force-Resisting System	ASCE 7 Section Where Detailing Requirements are Specified	Response Modification Coefficient, R^a	Overstrength Factor, Ω_0^g	Deflection Amplification Factor, C_d^b	Structural System Limitations including Structural Height, h_n (ft) Limits[c] Seismic Design Category				
					B	C	D[d]	E[d]	F[e]
E. DUAL SYSTEMS WITH INTERMEDIATE MOMENT FRAMES CAPABLE OF RESISTING AT LEAST 25% OF PRESCRIBED SEISMIC FORCES	12.2.5.1								
1. Steel special ~~steel~~ concentrically braced frames[f]	14.1	6	2 ½	5	NL	NL	35	NP	NP~~[h,k]~~
5. Steel and concrete composite ~~steel and concrete~~ special concentrically braced frames	14.3	5 ½	2 ½	4 ½	NL	NL	160	100	NP
6. Steel and concrete composite ordinary ~~composite~~ braced frames	14.3	3 ½	2 ½	3	NL	NL	NP	NP	NP
7. Steel and concrete composite ordinary ~~composite reinforced concrete~~ shear walls ~~with steel elements~~	14.3	5	3	4 ½	NL	NL	NP	NP	NP

Analysis and Commentary

The following describes the significant changes:

1. The material of construction now is almost always at the beginning of the description of a seismic force-resisting system. For instance, it is now "Steel special concentrically braced frames," rather than "Special steel concentrically braced frames."
2. Under Bearing Wall Systems and Building Frame Systems, "Light-frame walls sheathed with wood structural panels rated for shear resistance or steel sheets" are now divided into two items – one, wood; the other, cold-formed steel. The design coefficients and factors are not different for the two systems, but the referenced Chapter 14 section numbers (Column 2) are different. Also, "Light-frame wall systems using flat strap bracing" are now specifically indicated to be of cold-formed steel under Bearing Wall Systems.
3. Under Building Frame Systems, ASCE 7-05 included two different types of systems for both Eccentrically Braced Frames (EBF) and Buckling Restrained Braced Frames (BRBF). The primary distinction between these two subsystems was whether or not there were moment resisting beam-column connections within the braced bays. Testing at UC Berkeley (Uriz and Mahin, 2004) has indicated that designs that do not properly account for the stiffness and distribution of forces in braced frame connections may produce undesirable performance. ASCE 7-10 consolidates the EBF and BRBF building frame systems into a single designation, with proper consideration of the beam-column connection demands. The change allows the engineer to either provide a fully restrained moment connection meeting the requirements for ordinary moment connections in AISC 341, thereby directly providing a load path to resist the connection force and deformation demands, or to provide a connection with the ability to accommodate the potential rotation demands. An example of this would be a configuration tested at Lehigh University (Fahnestock et al., 2006), shown in Figure 1 of the cited reference that effectively formed a pinned condition in the beam just beyond the beam-column-brace connection.

References:
Uriz, P. and Mahin, S. A., "Seismic Performance Assessment of Concentrically Braced Steel Frames," *Proceedings* of the 13th World Conference on Earthquake Engineering; Vancouver, B.C., Canada, August 1-6, 2004, Paper No. 1639.

Fahnestock, L. A., Ricles, J. M., and Sause, R., "Experimental Study of a Large-Scale Buckling Restrained Frame Using the Psudo-Dynamic Testing Method," *Proceedings* of the 8th National Conference on Earthquake Engineering, San Francisco, CA, April, 2006.

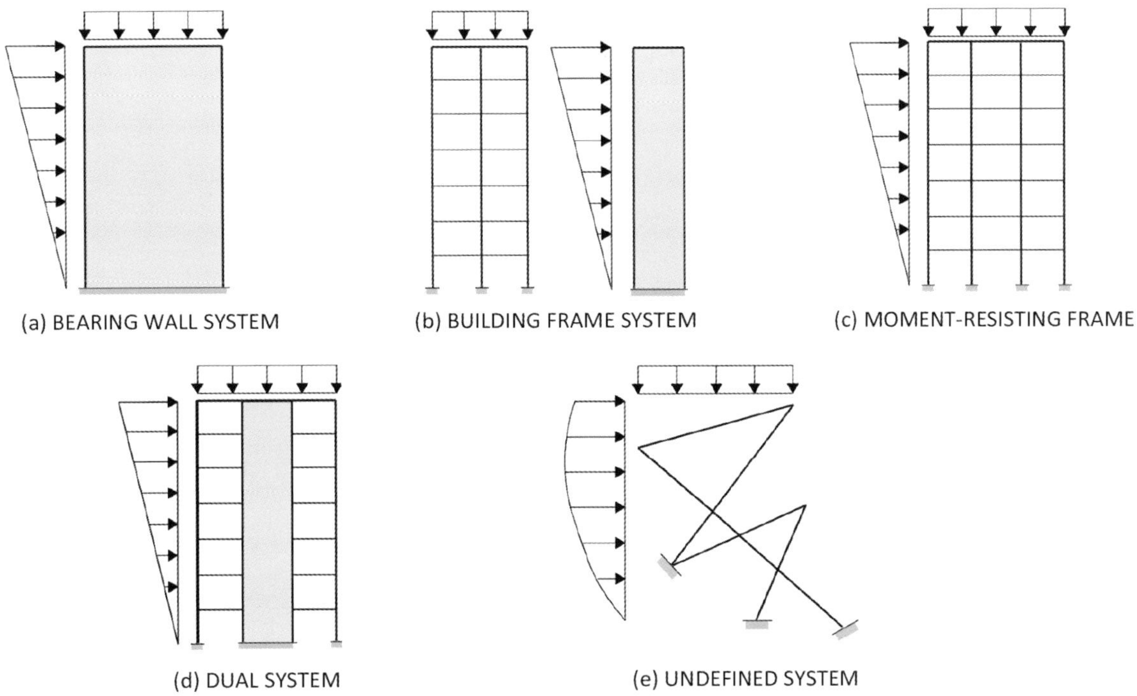

Seismic-Force-Resisting Structural Systems
Courtesy: International Code Council

Table 12.2-1 Addition

Design Coefficients and Factors for Autoclaved Aerated Concrete (AAC) Masonry Seismic Force-Resisting Systems

At a Glance

Bearing wall systems consisting of ordinary reinforced and ordinary plain autoclaved aerated concrete (AAC) masonry shear walls are added to Table 12.2-1.

2010 Standard

Table 12.2-1: Design Coefficients and Factors for Seismic Force-Resisting Systems

Seismic Force-Resisting System	ASCE 7 Section Where Detailing Requirements are Specified	Respone Modification Coefficient, R^a	Overstrength Factor, Ω_0^g	Deflection Amplification Factor, C_d^b	Structural System Limitations and Building Height, h_n (ft) Limits[c]				
					Seismic Design Category				
					B	C	D^d	E^d	F^e
A. Bearing Wall Systems									
13. Ordinary reinforced AAC masonry shear walls	14.4	2	2 ½	2	NL	35	NP	NP	NP
14. Ordinary plain AAC masonry shear walls	14.4	1 ½	2 ½	1 ½	NL	NP	NP	NP	NP

Analysis and Commentary

Seismic design factors and height restrictions for masonry structural systems made of autoclaved aerated concrete (AAC) are introduced in Table 12.2-1. The values and restrictions are consistent with those in 2009 IBC Section 1613.6.4.

The technical basis for AAC masonry shear wall systems includes system-level testing as well as component testing; it includes the precursor of today's ATC-63 approach (based on a 10% probability of collapse under MCE); and it includes comparison with values for similar current systems. That basis has been published in theses, dissertations, reports, and refereed journal publications. Three of the most accessible references are:

Tanner, J.E., Varela, J.L., and Klingner, R.E., "Design and Seismic Testing of a Two-story Full-scale Autoclaved Aerated Concrete (AAC) Assemblage Specimen," *Structural Journal*, American Concrete Institute, Farmington Hills, Michigan, Vol. 102, No. 1, January - February 2005, pp. 114-119.

Tanner, J.E., Varela, J.L., Klingner, R.E., Brightman M. J., and Cancino, U., "Seismic Testing of Autoclaved Aerated Concrete (AAC) Shear Walls: A Comprehensive Review," *Structural Journal*, American Concrete Institute, Farmington Hills, Michigan, Vol. 102, No. 3, May - June 2005, pp. 374-382.

Varela, J. L., Tanner, J. E., and Klingner, R. E., "Development of Seismic Force Reduction and Displacement Amplification Factors for Autoclaved Aerated Concrete Structures," Earthquake Spectra, Vol. 22, No.1, February 2006, pp. 267-286.

Laying AAC masonry units using thin-bed mortar and toothed trowel
www.masonrymagazine.com

An AAC masonry hotel in Las Palmas, Mexico, where AAC masonry elements are used as structural members as well as cladding
Courtesy: Autoclaved Aerated Concrete Products Association

Table 12.2-1 Addition

Seismic Design of Cold-Formed Steel – Special Bolted Moment Frames

At a Glance

A new moment-resisting frame system, cold-formed steel - special bolted moment frame, is added to Table 12.2-1 and modifications are made to the new referenced standard for the system.

2010 Standard

Table 12.2-1: Design Coefficients and Factors for Seismic Force-Resisting Systems

Seismic Force-Resisting System	ASCE 7 Section Where Detailing Requirements are Specified	Respone Modification Coefficient, R^a	Overstrength Factor, $\Omega_0{}^g$	Deflection Amplification Factor, $C_d{}^b$	Structural System Limitations and Building Height, h_n (ft) Limits[c]				
					Seismic Design Category				
					B	C	D[d]	E[d]	F[e]
C. Moment-Resisting Frame Systems									
12. Cold-formed steel – special bolted moment frame[p]	14.1	3 ½	3[o]	3.5	35	35	35	35	35

[o] Alternately, the seismic load effect with overstrength, E_{mh}, can be based on the expected strength determined in accordance with AISI S110.

[p] Cold-formed steel – special bolted moment frames shall be limited to one-story in height in accordance with AISI S110.

14.1.1 Reference Documents

The design, construction, and quality of steel members that resist seismic forces shall conform to the applicable requirements, as amended herein, of the following:

1. AISC 360
2. AISC 341
3. AISI S100
4. AISI S110
5. AISI S230
6. AISI S213
7. ASCE 19
8. ASCE 8
9. SJI-K-1.1
10. SJI-LH/DLH-1.1
11. SJI-JG-1.1
12. SJI-CJ-1.0

14.1.3 ~~14.1.2~~ Cold-Formed Steel

14.1.3.1 General

The design of cold-formed carbon or low-alloy steel structural members ~~to resist seismic loads~~ shall be in accordance with the requirements of AISI S100 and the design of cold-formed stainless steel structural members ~~to resist seismic loads~~ shall be in accordance with the requirements of ASCE 8. Where required, the seismic design of cold-formed steel structures shall be in accordance with the additional provisions of Section 14.1.3.2.

14.1.3.2 Seismic Requirements for Cold-Formed Steel Structures

Where a response modification coefficient, R, in accordance with Table 12.2-1 is used for the design of cold-formed steel structures, the structures shall be designed and detailed in accordance with the requirements of AISI S100, ASCE 8, and AISI S110 as modified in Section 14.1.3.3.

14.1.3.3 Modifications to AISI S110

The text of AISI S110 shall be modified as indicated in Sections 14.1.3.3.1 through 14.1.3.3.5. Italics are used for text within Sections 14.1.3.3.1 through 14.1.3.3.5 to indicate requirements that differ from AISI S110.

14.1.3.3.1 AISI S110, Section D1 Modify Section D1to read as follows:

D1 Cold-Formed Steel Special Bolted Moment Frames (CFS-SBMF)

Cold-formed steel–special bolted moment frame (CFS-SBMF) systems shall withstand significant inelastic deformations through friction and bearing at their bolted connections. Beams, columns, and connections shall satisfy the requirements in this section. CFS-SBMF systems shall be limited to one story structures, no greater than 35 feet in height, without column splices and satisfying the requirements in this section. *The CFS-SBMF shall engage all columns supporting the roof or floor above. The single size beam and single size column with the same bolted moment connection detail shall be used for each frame. The frame shall be supported on a level floor or foundation.*

14.1.3.3.2 AISI S110, Section D1.1.1 Modify Section D1.1.1 to read as follows:

D1.1.1 Connection Limitations

Beam-to-column connections in CFS-SBMF systems shall be bolted connections with snug-tight high-strength bolts. The bolt spacing and edge distance shall be in accordance with the limits of AISI S100, Section E3. *The 8-bolt configuration shown in Table D1-1 shall be used. The faying surfaces of the beam and column in the bolted moment connection region shall be free of lubricants or debris.*

14.1.3.3.3 AISI S110, Section D1.2.1 Modify Section D1.2.1 and add new Section D1.2.1.1 to read as follows:

D1.2.1 Beam Limitations

In addition to the requirements of Section D1.2.3, beams in CFS-SBMF systems shall be *ASTM A653 galvanized 55 ksi (374 MPa) yield stress* cold-formed *steel* C-section members with lips, and designed in accordance with Chapter C of AISI S100. *The beams shall have a minimum design thickness of 0.105 in. (2.67 mm). The beam depth shall be not less than 12 in. (305 mm) or greater than 20 in. (508 mm).* The flat depth-to-thickness ratio of the web shall not exceed $6.18\sqrt{E/F_y}$.

D1.2.1.1 Single-Channel Beam Limitations

When single-channel beams are used, torsional effects shall be accounted for in the design.

14.1.3.3.4 AISI S110, Section D1.2.2 Modify Section D1.2.2 to read as follows:

D1.2.2 Column Limitations

In addition to the requirements of D1.2.3, columns in CFS-SBMF systems shall be *ASTM A500 Grade B* cold-formed *steel* hollow structural section (HSS) members *painted with a standard industrial finished surface,* and designed in accordance with Chapter C of AISI S100. *The column depth shall be not less than 8 in. (203 mm) or greater than 12 in.(305 mm).* The flat depth-to-thickness ratio shall not exceed

$1.40\sqrt{E/F_y}$.

14.1.3.3.5 AISI S110, Section D1.3 Delete text in Section D1.3 to read as follows:

D1.3 Design Story Drift

Where the applicable building code does not contain design coefficients for CSF-SBMF systems, the provisions of Appendix 1 shall apply.

For structures having a period less than T_S, as defined in the applicable building code, alternate methods of computing Δ shall be permitted, provided such alternate methods are acceptable to the authority having jurisdiction.

Chapter 23
SEISMIC DESIGN REFERENCE DOCUMENTS

ANSI/AISI S110
Sections 14.1.1, 14.1.3.2, 14.1.3.3, Table 12.2-1
Standard for Seismic Design of Cold-Formed Steel Structural Systems – Special Bolted Moment Frames, 2007

Analysis and Commentary

A newly defined seismic force-resisting system entitled "Cold-formed Steel – Special Bolted Moment Frame" or CFS-SBMF is introduced in Table 12.2-1. Also, the first edition of AISI S110, *Standard for Seismic Design of Cold-Formed Steel Structural Systems – Special Bolted Moment Frames*, which is based upon research conducted by Uang and Sato at UCSD (2007), is adopted. The standard provides design provisions for the new system, which is expected to experience substantial inelastic deformations during significant seismic events. It is intended that most of the inelastic deformations will take place at the bolted connections, due to slip and bearing. In order to develop the designated mechanism, requirements based on capacity design principles are provided for the design of the beams, the columns, and the associated connections. Additionally, the document has specific requirements for the application of quality assurance and quality control procedures.

In Appendix 1, AISI S110 makes recommendations for the seismic design coefficients of the CFS-SBMF system. Cyclic testing has shown that CFS-SBMFs have very large ductility and significant hardening. Thus, R is set at 3.5. The derivation of the deflection amplification factor, C_d, can be found in AISI S110 Commentary Section D1.3. A capacity design procedure has been provided in Section D1.5 of AISI S110 Commentary such that the designer can explicitly calculate the seismic load effect with overstrength, E_{mh}, at the design story drift level. Alternatively, a conservative system overstrength factor, Ω_0, is also provided to be compatible with the conventional approach to computing E_{mh} in ASCE 7. Finally the height limitation of 35 ft for all SDCs is based on practical use only and not from any limits on the CFS-SBMF system strength.

The modifications to AISI S110 (2007 edition) are explained below:

- In Section 14.1.3.3.1, the language was modified to reflect that CFS-SBMF needs to use same-size beams and same-size columns throughout. In addition, the system needs to engage all primary columns, which support the roof or floor above, and those columns need to be supported on a level floor or foundation.

- In Section 14.1.3.3.2, the modifications were made for consistency with the test database.

- In Section 14.1.3.3.3, the modifications were made to be consistent with the test database (Uang and Sato, 2007); limitations on beam depth, steel grade, and surface treatment are added in Section D1.2.1 of AISI S110.

- In Section 14.1.3.3.4, the language was modified to be consistent with the test database (Uang and Sato, 2007); limitations on column depth, steel grade, and surface treatment are added in Section D1.2.2 of AISI S110. The width-thickness ratio was reduced based upon further review of the test specimens.

- As to Section 14.1.3.3.5, AISI S110 is intended primarily for industrial platforms; however, the standard is not limited to these non-building structures and does not prohibit architectural attachments (such as partition walls). Thus, the ASCE 7 Seismic Subcommittee requested that ASCE 7 Section 12.12 not be overwritten by AISI S110. Therefore, the $0.05h$ drift limit in Section D1.3 of AISI S110 has been eliminated in deference to the design story drift limits found in ASCE 7 Section 12.12. In addition, the first sentence in AISI S110 Section D1.3 was deleted because it was considered commentary.

Example of a Cold-Formed Steel – Special Bolted Moment Frame

12.2.3.1 Modification

R, C_d, and Ω_0 Values for Vertical Combinations

At a Glance

Where a structure has a vertical combination of structural systems in the same direction, the calculation of seismic design coefficients is clarified.

2010 Standard

*12.2.3.1 **R, C_d, and Ω_0 Values for Vertical Combinations***

Where a structure has a vertical combination in the same direction, the following requirements shall apply: ~~The value of the response modification coefficient, R, used for design at any story shall not exceed the lowest value of R that is used in the same direction at any story above that story. Likewise, the deflection amplification factor, C_d, and the system over strength factor, Ω_0, used for the design at any story shall not be less than the largest value of this factor that is used in the same direction at any story above that story.~~

1. Where the lower system has a lower Response Modification Coefficient, R, the design coefficients (R, Ω_0, and C_d) for the upper system are permitted to be used to calculate the forces and drifts of the upper system. For the design of the lower system, the design coefficients (R, Ω_0, and C_d) for the lower system shall be used. Forces transferred from the upper system to the lower system shall be increased by multiplying by the ratio of the higher response modification coefficient to the lower response modification coefficient.

2. Where the upper system has a lower Response Modification Coefficient, the Design Coefficients (R, Ω_0, and C_d) for the upper system shall be used for both systems.

(No change to exceptions)

Analysis and Commentary

This modification clarifies the calculation of seismic design coefficients where a lower portion of the structure has design coefficients that differ from those for the upper portion. The change looks almost like an editorial change. But it is in fact quite a substantive change, as indicated in the two figures on next page. When different lateral force-resisting systems are vertically stacked, the ASCE 7-05 rules concerning seismic design coefficients was that the R-value could not increase and that the values of Ω_0 and C_d could not decrease as one went down a building. According to ASCE 7-10, the R-value still cannot increase as one goes down a building. However, Ω_0 and C_d now must always correspond to the R-value.

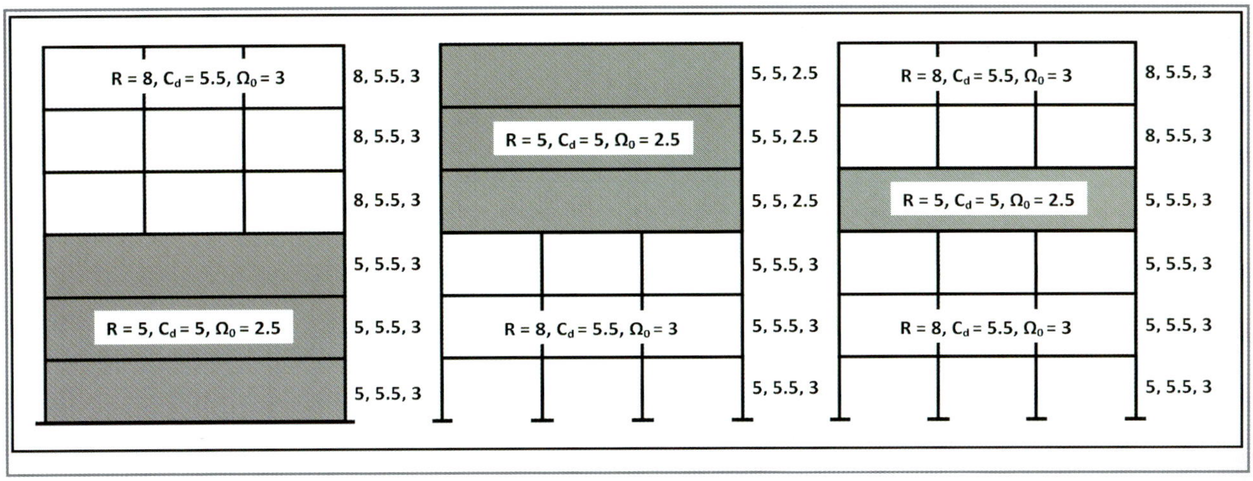

R: Cannot increase as you go down

C_d and Ω_0: Cannot decrease as you go down

ASCE 7-05 Vertical Combinations

Courtesy: S.K. Ghosh Associates Inc.

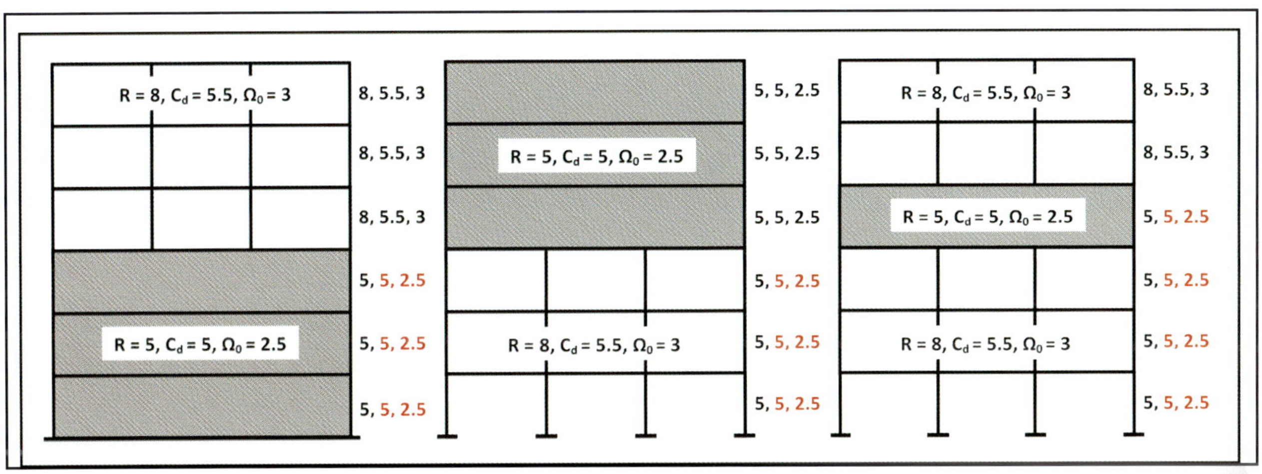

R: Cannot increase as you go down

C_d and Ω_0: Always correspond to R

ASCE 7-10 Vertical Combinations

Courtesy: S.K. Ghosh Associates Inc.

12.2.3.2 Modification

Two-Stage Analysis Procedure

At a Glance

Two-stage analysis procedure for structures having a flexible upper portion above a rigid lower portion is clarified.

2010 Standard

12.2.3.2 Two Stage Analysis Procedure

A two-stage equivalent lateral force procedure is permitted to be used for structures having a flexible upper portion above a rigid lower portion, provided ~~that~~ the design of the structure complies with all of the following:

a. The stiffness of the lower portion ~~must~~shall be at least 10 times the stiffness of the upper portion.

b. The period of the entire structure shall not be greater than 1.1 times the period of the upper portion considered as a separate structure ~~fixed~~ supported at the ~~base~~ transition from the upper to the lower portion.

c. The ~~flexible~~ upper portion shall be designed as a separate structure using the appropriate values of R and ρ.

d. The ~~rigid~~ lower portion shall be designed as a separate structure using the appropriate values of R and ρ. The reactions from the upper portion shall be those determined from the analysis of the upper portion amplified by the ratio of the R/ρ of the upper portion over R/ρ of the lower portion. This ratio shall not be less than 1.0.

e. The upper portion is analyzed with the equivalent lateral force or modal response spectrum procedure and the lower portion is analyzed with the equivalent lateral force procedure.

Analysis and Commentary

The location of the "base" in Condition (b) of the two-stage equivalent lateral force procedure is clarified. Section 11.2 defines "base" as the "level at which the horizontal seismic ground motions are considered to be imparted to the structure." Condition (b) of the two-stage equivalent lateral force procedure intends to reference the base of the upper portion of the structure, not the base of the entire structure. The definition of "base," however, applies to the entire structure.

Item e in Section 12.2.3.2 is added to clarify that a static or dynamic analysis can be performed on the upper portion and that a static analysis is to be performed on the lower portion. Since the lower portion is stiff, its seismic response will be dominated by the fundamental mode which makes equivalent static analysis the logical choice.

The other proposed changes are editorial. Adding "all of" in the charging language makes it clear that all of the conditions are required to be met. In Condition (a), "must" is changed to "shall" because "must" is not appropriate code language. In Conditions (c) and (d), "flexible" and "rigid" are deleted because they are redundant. The requirements for a flexible upper portion and a rigid lower portion are established in the charging language.

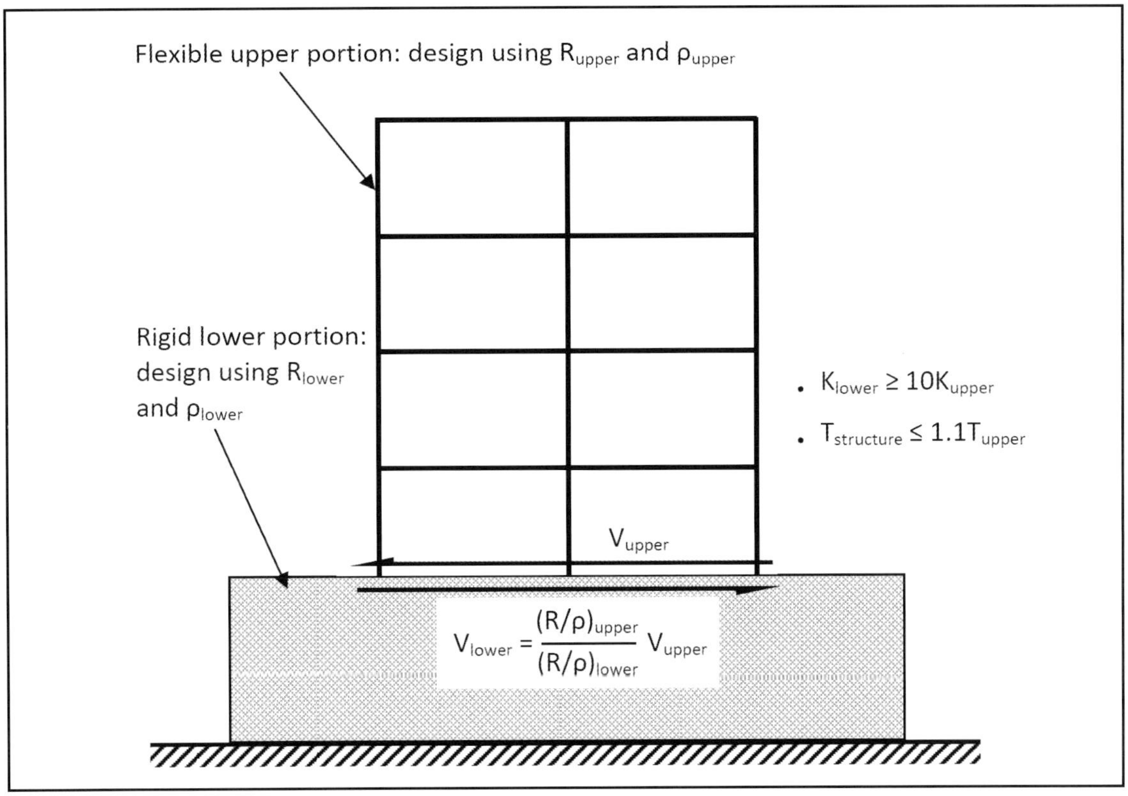

Conditions for Using Two Stage Analysis Procedure

12.2.5.2, Table 12.2-1

Modification and Clarification

Steel Cantilever Column Systems

At a Glance

Requirements for steel cantilever column systems are coordinated with the 2010 edition of AISC 341 and a footnote to Table 12.2-1 regarding the reduction of Ω_0 is clarified.

2010 Standard

12.2.5.2 Cantilever Column Systems

Cantilever column systems are permitted as indicated in Table 12.2-1 and as follows. The required axial strength of ~~axial load on~~ individual cantilever column elements, considering only the load combinations that include seismic load effects, ~~calculated in accordance with the load combinations of Section 2.3~~ shall not exceed 15 percent of the available axial strength ~~design strength of the column to resist axial loads alone~~, including slenderness effects. ~~or for allowable stress design, the axial load stress on individual cantilever column elements, calculated in accordance with the load combinations of Section 2.4 shall not exceed 15 percent of the permissible axial stress~~.

Foundation and other elements used to provide overturning resistance at the base of cantilever column elements shall be designed to resist the seismic load effects including overstrength factor of Section 12.4.3.

Table 12.2-1: Design Coefficients and Factors for Seismic Force-Resisting Systems[g]

Seismic Force-Resisting System	ASCE 7 Section Where Detailing Requirements are Specified	Respone Modification Coefficient, R^a	Overstrength Factor, Ω_0^g	Deflection Amplification Factor, C_d^b	Structural System Limitations and Building Height, h_n (ft) Limits[c]				
					Seismic Design Category				
					B	C	D[d]	E[d]	F[e]
G. CANTILEVERED COLUMN SYSTEMS DETAILED TO CONFORM TO THE REQUIREMENTS FOR:	12.2.5.2								
1. Steel special cantilever column systems~~moment frames~~	~~12.2.5.5 and~~ 14.1	2½	1¼	2½	35	35	35	35	35
~~2. Intermediate steel moment frames~~	~~14.1~~	~~1½~~	~~1¼~~	~~1½~~	~~35~~	~~35~~	~~35[h]~~	~~NP[h,i]~~	~~NP[h,i]~~
2 ~~3~~. Steel ordinary cantilever column systems ~~steel moment frames~~	14.1	1¼	1¼	1¼	35	35	NP[i]	NP[h,i]	NP[h,i]
3 ~~4~~. Special reinforced concrete moment frames[n]	12.2.5.5 and 14.2	2½	1¼	2½	35	35	35	35	35
4 ~~5~~. Intermediate reinforced concrete moment frames	14.2	1½	1¼	1½	35	35	NP	NP	NP
5 ~~6~~. Ordinary reinforced concrete moment frames	14.2	1	1¼	1	35	NP	NP	NP	NP
6 ~~7~~. Timber frames	14.5	1½	1½	1½	35	35	35	NP	NP
(No changes to remainder of table entries)									

[g]Where t~~T~~he tabulated value of the overstrength factor, Ω_0, is greater than or equal to 2 ½, Ω_0 is permitted to be reduced by subtracting the value of 1/2 ~~one-half~~ for structures with flexible diaphragms~~, but shall not be taken as less than 2.0 for any structure except cantilever column systems.~~

Analysis and Commentary

ASCE 7-05 contained provisions for steel ordinary, intermediate as well as special cantilever column systems. In previous editions, AISC 341 did not explicitly address cantilever column systems. Consequently, the resulting set of requirements associated with each system was vague, confusing, and potentially incomplete.

Modifications made to Section 12.2.5.2 and Table 12.2-1 have been coordinated with parallel changes in the 2010 edition of AISC 341. The completed edition of AISC 341-10 contains specific requirements for just two cantilevered column systems – ordinary and special. The special cantilevered column system utilizes the appropriate compactness and strength requirements for special moment frame columns and their connections. However, stability bracing of the columns is permitted to meet the requirements for moderately ductile members. This is believed to be reasonable given the already low *R*-factor assigned for these systems; the use of amplified forces in their design; and, the fact that stability bracing has never been required before on these types of systems. The ordinary cantilevered column system requirements are based on those for ordinary moment frame columns and their connections. AISC 341-10 does not have separate requirements for intermediate cantilevered column systems.

The reduction in the overstrength factor, Ω_0, permitted by footnote *g* of Table 12.2-1, is clarified. Neither the reduction by subtracting ½ nor the 2.0 limit applies to cantilevered column systems, for which the value of Ω_0 is 1¼ or 1½. Also, the words "one-half" were confusing and could be construed to erroneously mean one half of Ω_0 rather than the value of ½.

Gas station canopies are typically cantilevered column systems.

12.2.5.4

Height Limit for Special Steel Plate Shear Walls

Modification and Addition

At a Glance

Increased building height limits for steel braced frames and special reinforced concrete shear walls is extended to Steel Special Plate Shear Walls

2010 Standard

12.2.5.4 Increased Structural Height Limit for Steel Eccentrically Braced Frames, Steel Special Concentrically Braced Frames, Steel Buckling-restrained Braced Frames, Steel Special Plate Shear Walls and Special Reinforced Concrete Shear Walls

The limits on structural height, h_n, in Table 12.2-1 are permitted to be increased from 160 ft (50 m) to 240 ft (75 m) for structures assigned to Seismic Design Categories D or E and from 100 ft (30 m) to 160 ft (50 m) for structures assigned to Seismic Design Category F provided the seismic force-resisting systems are limited to steel eccentrically braced frames, steel special concentrically braced frames, steel buckling-restrained braced frames, steel special plate shear walls or special reinforced concrete cast-in-place shear walls and ~~that meet~~ both of the following requirements are met :

1. The structure shall not have an extreme torsional irregularity as defined in Table 12.2-1 (horizontal structural irregularity Type 1b).
2. The steel eccentrically braced frames, steel special concentrically braced frames, steel buckling restrained braced frames, steel special plate shear walls or special reinforced cast-in-place concrete shear walls in any one plane shall resist no more than 60 percent of the total seismic forces in each direction, neglecting accidental torsional effects.

Analysis and Commentary

Steel Special Plate Shear Wall systems were first introduced in the 2005 editions of ASCE 7 and AISC 341. During the incorporation of the seismic design parameters and height limitations for the system into ASCE 7-05 Table 12.2-1, the inclusion of this system in the permitted height increase of ASCE 7-05 Section 12.2.5.4 was overlooked. This modification includes these systems in the permitted height increase of ASCE 7-10 Section 12.2.5.4.

The other changes are made so that terminology in Section 12.2.5.4 corresponds fully to that in Table 12.2-1.

Special Steel Plate Shear Walls

Courtesy: American Institute of Steel Construction (AISC)

12.2.5.6, 12.2.5.6.1, 12.2.5.6.2, 12.2.5.7, 12.2.5.7.1, 12.2.5.7.2, 12.2.5.7.3, 12.2.5.8

Modification and Addition

Steel Ordinary Moment Frames and Steel Intermediate Moment Frames

At a Glance

Provisions concerning restrictions on ordinary and intermediate moment frames in high seismic design categories are reorganized. Exceptions are added for steel industrial buildings.

2010 Standard

~~**12.2.5.6 Single-Story Steel Ordinary and Intermediate Moment Frames in Structures Assigned to Seismic Design Category D or E.** Single-story steel ordinary moment frames and intermediate moment frames in structures assigned to Seismic Design Category D or E are permitted up to a height of 65 ft (20 m) where the dead load supported by and tributary to the roof does not exceed 20 psf (0.96 kN/m²). In addition, the dead loads tributary to the moment frame, of the exterior wall more than 35 ft above the base shall not exceed 20 psf (0.96 kN/m²).~~

~~**12.2.5.7 Other Steel Ordinary and Intermediate Moment Frames in Structures Assigned to Seismic Design Category D or E.** Steel ordinary moment frames in structures assigned to Seismic Design Category D or E not meeting the limitations set forth in Section 12.2.5.6 are permitted within light-frame construction up to a height of 35 ft (10.6 m) where neither the roof nor the floor dead load supported by and tributary to the moment frames exceeds 35 psf (1.68 kN/m²). In addition, the dead load of the exterior walls tributary to the moment frame shall not exceed 20 psf (0.96 kN/m²). Steel intermediate moment frames in structures assigned to Seismic Design Category D or E not meeting the limitations set forth in Section 12.2.5.6 are permitted as follows:~~

1. ~~In Seismic Design Category D, intermediate moment frames are permitted to a height of 35 ft (10.6 m).~~
2. ~~In Seismic Design Category E, intermediate moment frames are permitted to a l height of 35 ft (10.6 m) provided neither the roof nor the floor dead load supported by and tributary to the moment frames exceeds 35 psf (1.68 kN/m²). In addition, the dead load of the exterior walls tributary to the moment frame shall not exceed 20 psf (0.96 kN/m²).~~

~~**12.2.5.8 Single-Story Steel Ordinary and Intermediate Moment Frames in Structures Assigned to Seismic Design Category F.** Single-story steel ordinary moment frames and intermediate moment frames in structures assigned to Seismic Design Category F are permitted up to a height of 65 ft (20 m) where the dead load supported by and tributary to the roof does not exceed 20 psf (0.96 kN/m²). In addition, the dead loads of the exterior walls tributary to the moment frame shall not exceed 20 psf (0.96 kN/m²).~~

~~**12.2.5.9 Other Steel Intermediate Moment Frame Limitations in Structures Assigned to Seismic Design Category F.** In addition to the limitations for steel intermediate moment frames in structures assigned to Seismic Design Category E as set forth in Section 12.2.5.7, steel intermediate moment frames in structures assigned to Seismic Design Category F are permitted in light-frame construction.~~

12.2.5.6 Steel Ordinary Moment Frames

12.2.5.6.1 Seismic Design Category D or E

a. Single-story steel ordinary moment frames in structures assigned to Seismic Design Category D or E are permitted up to a structural height, h_n, of 65 ft (20 m) where the dead load supported by and tributary to the roof does not exceed 20 psf (0.96 kN/m^2). In addition, the dead load of the exterior walls more than 35 ft (10.6 m) above the base tributary to the moment frames shall not exceed 20 psf (0.96 kN/m^2).

EXCEPTION: Single-story structures with steel ordinary moment frames or intermediate moment frames whose purpose is to enclose equipment or machinery and whose occupants are engaged in maintenance or monitoring of that equipment, machinery, or their associated processes shall be permitted to be of unlimited height where the sum of the dead and equipment loads supported by and tributary to the roof does not exceed 20 psf (0.96 kN/m^2). In addition, the dead load of the exterior wall system including exterior columns more than 35 ft (10.6 m) above the base shall not exceed 20 psf (0.96 kN/m^2). For determining compliance with the exterior wall or roof load limits, the weight of equipment or machinery, including cranes, not self-supporting for all loads shall be assumed fully tributary to the area of the adjacent exterior wall or roof not to exceed 600 ft^2 (55.8 m^2) regardless of their height above the base of the structure.

b. Steel ordinary moment frames in structures assigned to Seismic Design Category D or E not meeting the limitations set forth in Section 12.2.5.6.1.a are permitted within light-frame construction up to a structural height, h_n, of 35 ft (10.6 m) where neither the roof dead load nor the dead load of any floor above the base supported by and tributary to the moment frames exceeds 35 psf (1.68 kN/m^2). In addition, the dead load of the exterior walls tributary to the moment frames shall not exceed 20 psf (0.96 kN/m^2).

12.2.5.6.2 Seismic Design Category F. Single-story steel ordinary moment frames in structures assigned to Seismic Design Category F are permitted up to a structural height, h_n, of 65 ft (20 m) where the dead load supported by and tributary to the roof does not exceed 20 psf (0.96 kN/m^2). In addition, the dead load of the exterior walls tributary to the moment frames shall not exceed 20 psf (0.96 kN/m^2).

12.2.5.7 Steel Intermediate Moment Frames

12.2.5.7.1 Seismic Design Category D

a. Single-story steel intermediate moment frames in structures assigned to Seismic Design Category D are permitted up to a structural height, h_n, of 65 ft (20 m) where the dead load supported by and tributary to the roof does not exceed 20 psf (0.96 kN/m^2). In addition, the dead load of the exterior walls more than 35 ft (10.6 m) above the base tributary to the moment frames shall not exceed 20 psf (0.96 kN/m^2).

EXCEPTION: Single-story structures with steel intermediate moment frames whose purpose is to enclose equipment or machinery, and whose occupants are engaged in maintenance or monitoring of that equipment, machinery, or their associated processes shall be permitted to be of unlimited height where the sum of the dead and equipment loads supported by and tributary to the roof does not exceed 20 psf (0.96 kN/m^2). In addition, the dead load of the exterior wall system including exterior columns more than 35 ft (10.6 m) above the base shall not exceed 20 psf (0.96 kN/m^2). For determining compliance with the exterior wall or roof load limits, the weight of equipment or machinery, including cranes, not self-supporting for all loads shall be assumed fully tributary to an area of the adjacent exterior wall or roof not to exceed 600 ft^2 (55.8 m^2) regardless of their height above the base of the structure.

b. Steel intermediate moment frames in structures assigned to Seismic Design Category D not meeting the limitations set forth in Section 12.2.5.7.1.a are permitted to a structural height, h_n, of 35 ft (10.6 m).

12.2.5.7.2 Seismic Design Category E

a. Single-story steel intermediate moment frames in structures assigned to Seismic Design Category E are permitted up to a structural height, h_n, of 65 ft (20 m) where the dead load supported by and tributary to the roof does not exceed 20 psf (0.96 kN/m^2). In addition, the dead load of the exterior walls more than 35 ft (10.6 m) above the base tributary to the moment frames shall not exceed 20 psf (0.96 kN/m^2).

EXCEPTION: Single story structures with steel intermediate moment frames whose purpose is to enclose equipment or machinery and whose occupants are engaged in maintenance or monitoring of that equipment, machinery, or their associated processes shall be permitted to be of unlimited height where the sum of the dead and equipment loads supported by and tributary to the roof does not exceed 20 psf (0.96 kN/m^2). In addition, the dead load of the exterior wall system including exterior columns more than 35 ft (10.6 m) above the base shall not exceed 20 psf (0.96 kN/m^2). For determining compliance with the exterior wall or roof load limits, the weight of equipment or machinery, including cranes, not self-supporting for all loads shall be assumed fully tributary to the area of the adjacent exterior wall or roof not exceeding 600 ft^2 (55.8 m^2) regardless of their height above the base of the structure.

b. Steel intermediate moment frames in structures assigned to Seismic Design Category E not meeting the limitations set forth in Section 12.2.5.7.2.a are permitted to a structural height, h_n, of 35 ft (10.6 m) where neither the roof dead load nor the dead load of any floor above the base supported by and tributary to the moment frames exceeds 35 psf (1.68 kN/m^2). In addition, the dead load of the exterior walls tributary to the moment frames shall not exceed 20 psf (0.96 kN/m^2).

12.2.5.7.3 Seismic Design Category F

a. Single-story steel intermediate moment frames in structures assigned to Seismic Design Category F are permitted up to a structural height, h_n, of 65 ft (20 m) where the dead load supported by and tributary to the roof does not exceed 20 psf (0.96 kN/m^2). In addition, the dead load of the exterior walls tributary to the moment frames shall not exceed 20 psf (0.96 kN/m^2).

b. Steel intermediate moment frames in structures assigned to Seismic Design Category F not meeting the limitations set forth in Section 12.2.5.7.3.a are permitted within light-frame construction up to a structural height, h_n, of 35 ft (10.6 m) where neither the roof dead load nor the dead load of any floor above the base supported by and tributary to the moment frames exceeds 35 psf (1.68 kN/m^2). In addition, the dead load of the exterior walls tributary to the moment frames shall not exceed 20 psf (0.96 kN/m^2).

12.2.5.8 ***~~12.2.5.10~~* *Shear Wall-Frame Interactive Systems***

The shear strength of the shear walls of the shear wall-frame interactive system shall be at least 75 percent of the design story shear at each story. The frames of the shear wall-frame interactive system shall be capable of resisting at least 25 percent of the design story shear in every story.

Analysis and Commentary

ASCE 7-05 Sections 12.2.5.6 through 12.2.5.9 are reorganized in a significant way to remove inconsistencies and to improve readability. The new organization follows the simpler organization of Table 12.2-1, starting with the seismic force-resisting systems followed by Seismic Design Categories.

In addition, exceptions are added. These exceptions were developed at an AISC- sponsored Industrial Building and Seismic Design Summit Meeting held in January 2009. The purpose of the meeting was to find safe and suitable solutions to address the problems industrial buildings had in meeting the system and height limitations in ASCE 7-05 Table 12.2-1.

Ordinary steel moment-frame construction has been used for many years for tall, single-story buildings used in a variety of applications including mill buildings, aircraft maintenance and assembly structures, and similar applications. The performance of these structures in past strong motion earthquakes has reportedly been satisfactory.

Despite the repeated good performance of such structures, ASCE 7-05 prohibited the use of ordinary and intermediate moment frames in higher seismic design categories for many of these structures. New exceptions are added for SDC D and E ordinary and intermediate moment frames to allow this group of structures that has historically performed well.

The following important items are worth noting:

1. To allow unlimited height, the sum of the dead and equipment load cannot be greater than 20 psf.
2. The exterior wall weight must include the weight of exterior columns.
3. For the case when cranes or other equipment are not self-supporting for all loads (i.e., supported for vertical loads and/or laterally braced by columns that are part of or stabilized by OMF/IMF frames), the operating weight must be treated as fully tributary (100%) to either the adjacent exterior wall when located in an exterior bay or to the adjacent roof when located in an interior bay. The tributary area used for weight distribution must not exceed 600 square feet (weights in exterior bays can also be tributary to the roof if desired).

Crane self-supporting for vertical loads but not self-supporting for lateral loads

Crane not self-supporting for vertical and lateral loads

12.3.1.1 Addition

Flexible Diaphragm Condition

At a Glance

A new prescriptive condition is added for flexible diaphragm classification for light-frame construction.

2010 Standard

~~**12.3.1.1 Flexible Diaphragm Condition.** Diaphragms constructed of untopped steel decking or wood structural panels are permitted to be idealized as flexible in structures in which the vertical elements are steel or composite steel and concrete braced frames, or concrete, masonry, steel or composite shear walls. Diaphragms of wood structural panels or untopped steel decks in one and two-family residential buildings of light-frame construction shall also be permitted to be idealized as flexible.~~

12.3.1.1 Flexible Diaphragm Condition

Diaphragms constructed of untopped steel decking or wood structural panels are permitted to be idealized as flexible if any of the following conditions exist:

a. In structures where the vertical elements are steel braced frames, steel and concrete composite braced frames or concrete, masonry, steel, or steel and concrete composite shear walls.

b. In one- and two-family dwellings.

c. In structures of light-frame construction where all of the following conditions are met:

   1. Topping of concrete or similar materials is not placed over wood structural panel diaphragms except for nonstructural topping no greater than 1½ in. (38 mm) thick.

   2. Each line of vertical elements of the seismic force-resisting system complies with the allowable story drift of Table 12.12-1.

Analysis and Commentary

ASCE 7-05 Section 12.3.1.1 set forth conditions under which certain diaphragms may be considered flexible for the purposes of lateral force distribution. 2006 IBC (Section 1613.6.1) modified this ASCE 7-05 section to add one set of other conditions, the satisfaction of which would qualify a diaphragm as flexible. This modification was continued in the 2009 IBC. An abbreviated set of the conditions included in this IBC modification is now part of ASCE 7-10.

The additional set of conditions is based on results from the CUREE-Caltech Woodframe Project which showed that for regular, light-framed, wood diaphragm buildings, treating the diaphragms as flexible gives a better match with full size experimental tests. The CUREE results further demonstrated that designing shear wall lines based on tributary area forces more stiffness into wall lines composed of narrow panels, reducing the eccentric torsion. Designing by rigid-diaphragm analysis tends to transfer forces away from light wall lines, allowing them to be even lighter, thus exacerbating an undesired condition. There is also analytical evidence that typical light-framed wood structures are more accurately modeled using the flexible diaphragm assumption.

CUREE has shown by full-size tests that a thin, lightweight, nonstructural cellular concrete or gypsum topping does not appreciably change the stiffness of a wood diaphragm.

Requiring separate shear wall lines to meet the drift criterion is a CUREE recommendation. This ensures that the vertical elements of the lateral force-resisting system are substantial enough to share load on a tributary basis and not require incompatibility of deflection, which might result in problematic torsion. One example is an open-sided building with fixed-base pipe columns along the open face. The pipe columns were designed for tributary load, but were too flexible to receive the load. The building diaphragm rotated excessively, causing failure. It would be best to address this concern directly by considering torsion and deflection incompatibility while preserving the flexible diaphragm assumption, which produces more accurate results of analysis for these buildings.

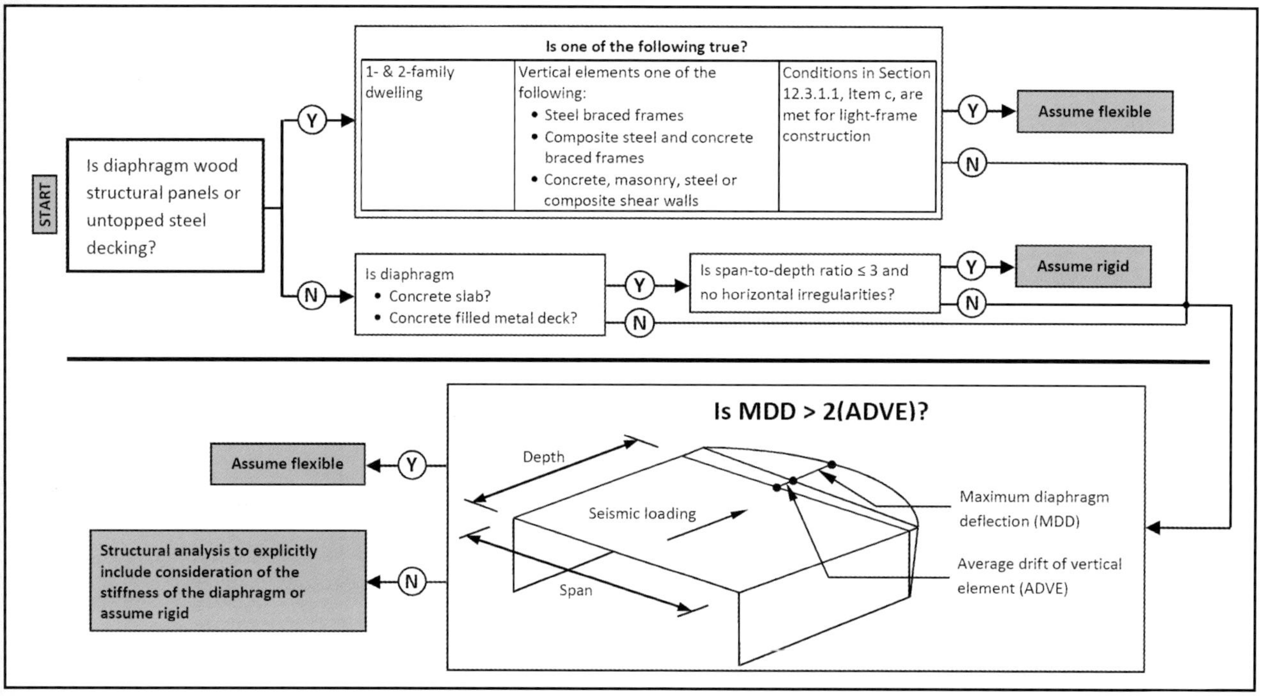

Flowchart for Determining Diaphragm Flexibility
Courtesy: S.K. Ghosh Associates Inc.

Table 12.3-1 Modification

Horizontal Structural Irregularities

At a Glance

The definitions of a number of horizontal irregularities in Table 12.3-1 are modified – one of them substantially.

2010 Standard

Table 12.3-1 Horizontal Structural Irregularities

Type	Description	Reference Section	Seismic Design Category Application
1a.	**Torsional Irregularity:** Torsional irregularity is defined to exist where the maximum story drift, computed including accidental torsion with $A_x = 1.0$, at one end of the structure transverse to an axis is more than 1.2 times the average of the story drifts at the two ends of the structure. Torsional irregularity requirements in the reference sections apply only to structures in which the diaphragms are rigid or semi-rigid.	12.3.3.4 12.7.3 12.8.4.3 12.12.1 Table 12.6-1 Section 16.2.2	D, E, and F B, C, D, E, and F C, D, E, and F C, D, E, and F D, E, and F B, C, D, E, and F
1b.	**Extreme Torsional Irregularity:** Extreme torsional irregularity is defined to exist where the maximum story drift, computed including accidental torsion with $A_x = 1.0$, at one end of the structure transverse to an axis is more than 1.4 times the average of the story drifts at the two ends of the structure. Extreme torsional irregularity requirements in the reference sections apply only to structures in which the diaphragms are rigid or semi-rigid.	12.3.3.1 12.3.3.4 12.7.3 12.8.4.3 12.12.1 Table 12.6-1 Section 16.2.2	E and F D B, C, and D C and D C and D D B, C, and D
2.	**Re-entrant Corner Irregularity:** Re-entrant corner irregularity is defined to exist where both plan projections of the structure beyond a re-entrant corner are greater than 15% of the plan dimension of the structure in the given direction.	12.3.3.4 Table 12.6-1	D, E, and F D, E, and F
3.	**Diaphragm Discontinuity Irregularity:** Diaphragm discontinuity irregularity is defined to exist where there is a diaphragm ~~are diaphragms~~ with an abrupt discontinuity ~~discontinuities~~ or variation~~s~~ in stiffness, including one ~~those~~ having a cutout or open area~~s~~ greater than 50% of the gross enclosed diaphragm area, or a change~~s~~ in effective diaphragm stiffness of more than 50% from one story to the next.	12.3.3.4 Table 12.6-1	D, E, and F D, E, and F
4.	**Out-of-Plane Offset~~s~~ Irregularity:** Out-of-plane offset irregularity is defined to exist where there is a discontinuity ~~are discontinuities~~ in a lateral force-resistance path, such as an out-of-plane offset~~s~~ of at least one of the vertical elements.	12.3.3.3 12.3.3.4 12.7.3 Table 12.6-1 Section 16.2.2	B, C, D, E, and F D, E, and F B, C, D, E, and F D, E, and F B, C, D, E, and F
5.	**Nonparallel System~~s~~ Irregularity:** Nonparallel system irregularity is defined to exist where ~~the~~ vertical lateral force-resisting elements are not parallel to ~~or symmetric about~~ the major orthogonal axes of the seismic force-resisting system.	12.5.3 12.7.3 Table 12.6-1 Section 16.2.2	C, D, E, and F B, C, D, E, and F D, E, and F B, C, D, E, and F

Analysis and Commentary

In Section 12.8.4.3 the calculation of the torsional amplification factor, A_x, is not iterative. A_x is calculated from an analysis performed assuming $A_x = 1.0$. Similarly, the classification of torsional irregularity should not be iterative.

Items 3, 4, and 5 are revised editorially to switch from plural to singular.

The text of item 5 clearly indicates that the nonparallel system irregularity exists only where the vertical elements are not parallel to the major orthogonal axes. In other words, being parallel to the major orthogonal axes is sufficient to eliminate the irregularity. The previous text of "parallel to or symmetric about" was sometimes misread to require that the system be both parallel to *and* symmetric about the major orthogonal axes. By that reading, Figure (b) below has a nonparallel system irregularity. While it may be possible to recast the text to improve clarity, the analysis and design requirements for systems with horizontal irregularity 5 (namely, 3D analysis and consideration of orthogonal effects) are such that application to all systems with nonparallel systems (even if the systems are symmetric) is not burdensome. With this change, a system with a plan as in Figure (a) would require special treatment, but a system with a plan as in Figure (b) would not.

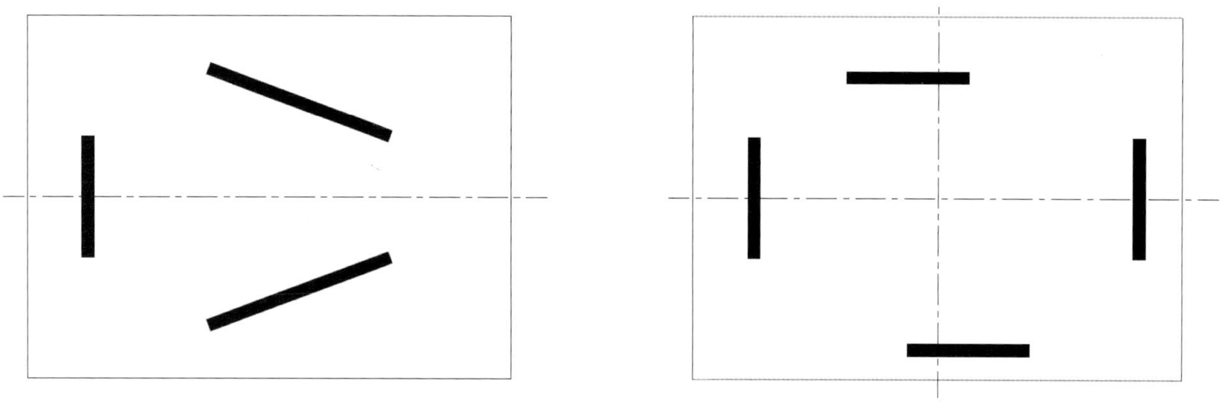

Figure (a) **Figure (b)**

Table 12.3-2 Modification

Vertical Structural Irregularities

At a Glance

The definition for In-Plane Discontinuity in Vertical Lateral Force-Resisting Element Irregularity is revised.

2010 Standard

Table 12.3-2 Vertical Structural Irregularities

Type	Description	Reference Section	Seismic Design Category Application
(No change to table for Types 1a -3)			
4.	**In-Plane Discontinuity in Vertical Lateral Force-Resisting Element Irregularity:** In-plane discontinuity in vertical lateral force-resisting element irregularity is defined to exist where there is an in-plane offset of a vertical seismic ~~the lateral~~ force-resisting element resulting in overturning demands on a supporting beam, column, truss, or slab. ~~greater than the length of those elements or there exists a reduction in stiffness of the resisting element in the story below.~~	12.3.3.3 12.3.3.4 Table 12.6-1	B, C, D, E, and F D, E, and F D, E, and F
(No change to table for Types 5a and 5b)			

Analysis and Commentary

The definition for Vertical Structural Irregularity Type 4, In-Plane Discontinuity in Vertical Lateral Force-Resisting Element Irregularity, is revised.

ASCE 7-05 required an in-plane offset to be greater than the length of an element for the requirements of Section 12.3.3.3 to apply. In Figure (a), the column marked with a circle (below the discontinued shear wall) would have no special design requirements per ASCE 7-05; that condition is unsafe. New language is now provided to remedy this situation.

There are many cases in which a lateral force-resisting element may have a reduction in stiffness in the story below without causing an in-plane discontinuity. The moment frame systems illustrated in Figures (b) and (c) and the shear wall system illustrated in Figure (d) would all be classified as having an in-plane discontinuity by ASCE 7-05 text. However, systems like these are common and can perform well.

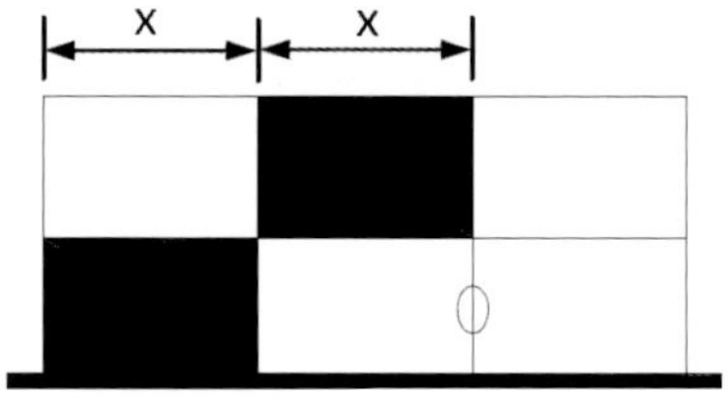

Figure (a) Does not qualify as Vertical Irregularity 4

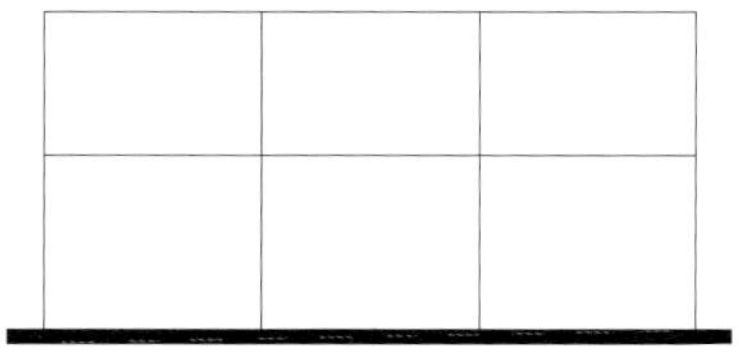

Figure (b) Reduction in stiffness due to increase in column height

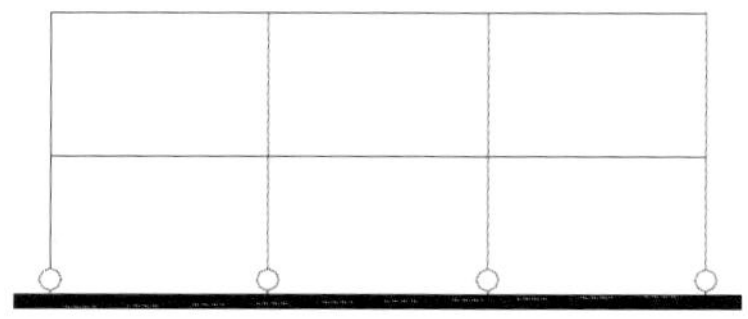

Figure (c) Reduction in stiffness due to pinned connection at base

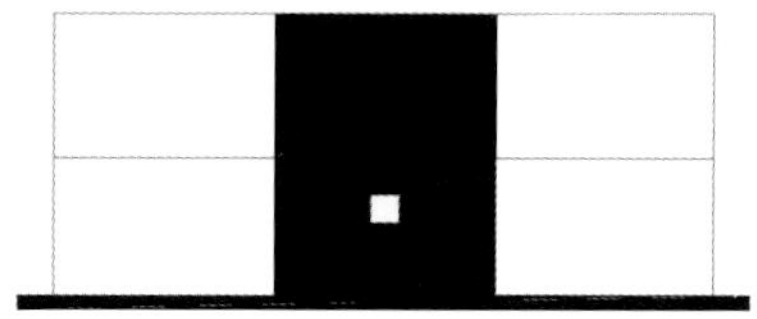

Figure (d) Reduction in stiffness due to opening in shear wall

Table 12.3-3 Clarification

Redundancy Provisions

At a Glance

The definition of height-to-length ratio of shear walls is clarified for determination of redundancy coefficient.

2010 Standard

Table 12.3-3 Requirements for Each Story Resisting More Than 35% of the Base Shear

Lateral Force-Resisting Element	Requirement
Braced Frames	*(no change to text)*
Moment Frames	*(no change to text)*
Shear Walls or Wall Piers with a height-to-length ratio ~~of~~ greater than 1.0	Removal of a shear wall or wall pier with a height-to-length ratio greater than 1.0 within any story, or collector connections thereto, would not result in more than a 33% reduction in story strength, nor does the resulting system have an extreme torsional irregularity (horizontal structural irregularity Type 1b). The shear wall and wall pier height-to-length ratios are determined as shown in Figure 12.3-2.
Cantilever Columns	*(no change to text)*
Other	No requirements

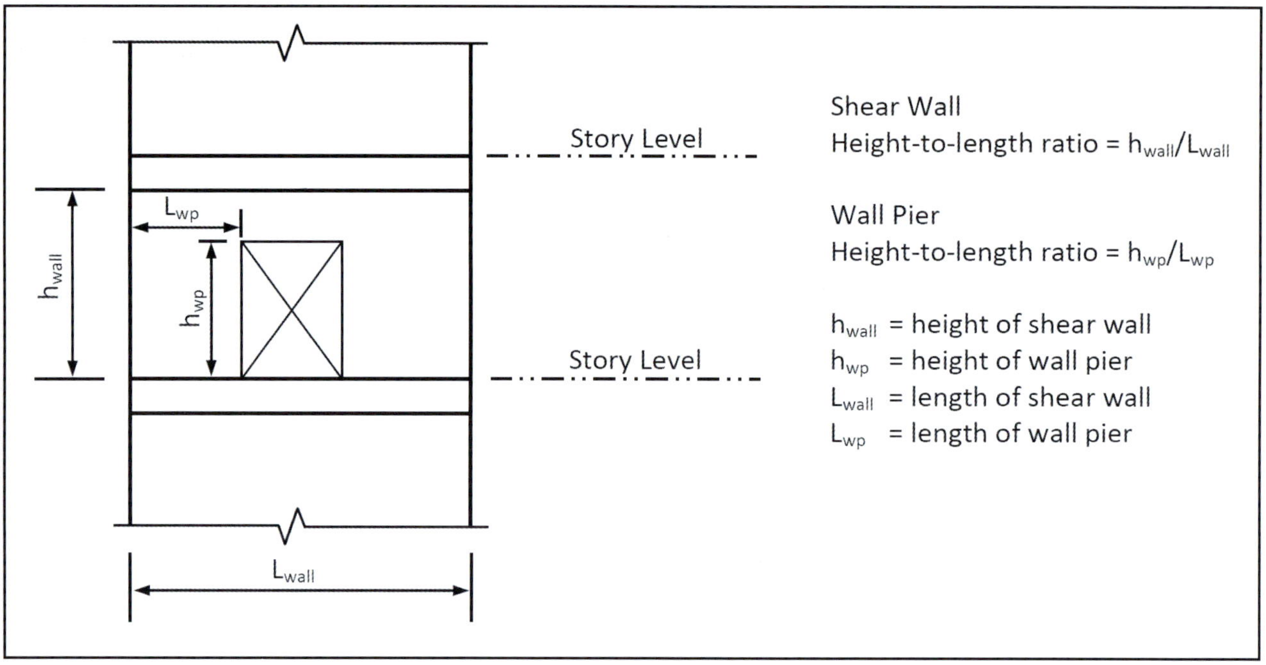

<u>**FIGURE 12.3-2 Shear Wall and Wall Pier Height-to-Length Ratio Determination**</u>

Analysis and Commentary

This change clarifies the definition of height-to-length ratio for shear walls and wall piers for the purpose of determining the redundancy factor, ρ. Wall height is from the top of a floor to the underside of the horizontal framing for the floor above, rather than to the top of the floor above. 4-ft-long plywood shear walls are thus sufficient to produce a redundancy factor of one for top-of-floor to top-of-floor height exceeding 8 ft, provided the wall height shown in Figure 12.3-2 does not exceed 8 ft.

New Figure 12.3-2 will help code users determine height-to-length ratio for shear walls and wall piers.

12.3.3.4 Modification

Increase in Forces Due to Irregularities for Seismic Design Categories D through F

At a Glance

The provision concerning increase in forces due to irregularities for Seismic Design Categories D through F is simplified and clarified; an error is corrected.

2010 Standard

12.3.3.4 Increase in Forces Due to Irregularities for Seismic Design Categories D through F
For structures assigned to Seismic Design Category D, E, or F and having a horizontal structural irregularity of Type 1a, 1b, 2, 3, or 4 in Table 12.3-1 or a vertical structural irregularity of Type 4 in Table 12.3-2, the design forces determined from Section ~~12.8.1~~ 12.10.1.1 shall be increased 25 percent for the following elements of the seismic force-resisting system:

1. ~~C~~connections of diaphragms to vertical elements and to collectors.
2. Collectors and their connections, including connections to vertical elements, of the seismic force-resisting system. ~~and to connections of collectors to the vertical elements.~~

~~Collectors and their connections also shall be designed for these increased forces unless they are designed for the load combinations with overstrength factor of Section 12.4.3.2, in accordance with Section 12.10.2.1.~~

EXCEPTION:

Forces calculated using the seismic load effects including overstrength factor of Section 12.4.3 need not be increased.

Analysis and Commentary

The ASCE 7-05 requirement concerning increase in forces due to irregularities for Seismic Design Categories D through F is simplified by presenting the exception as such. The proposal also corrects the incorrect reference to the equivalent lateral force base shear in ASCE 7-05 Section 12.8.1 (and, by implication, the corresponding vertical distribution) and refers to the diaphragm design force in ASCE 7-10 Section 12.10.1.1 instead.

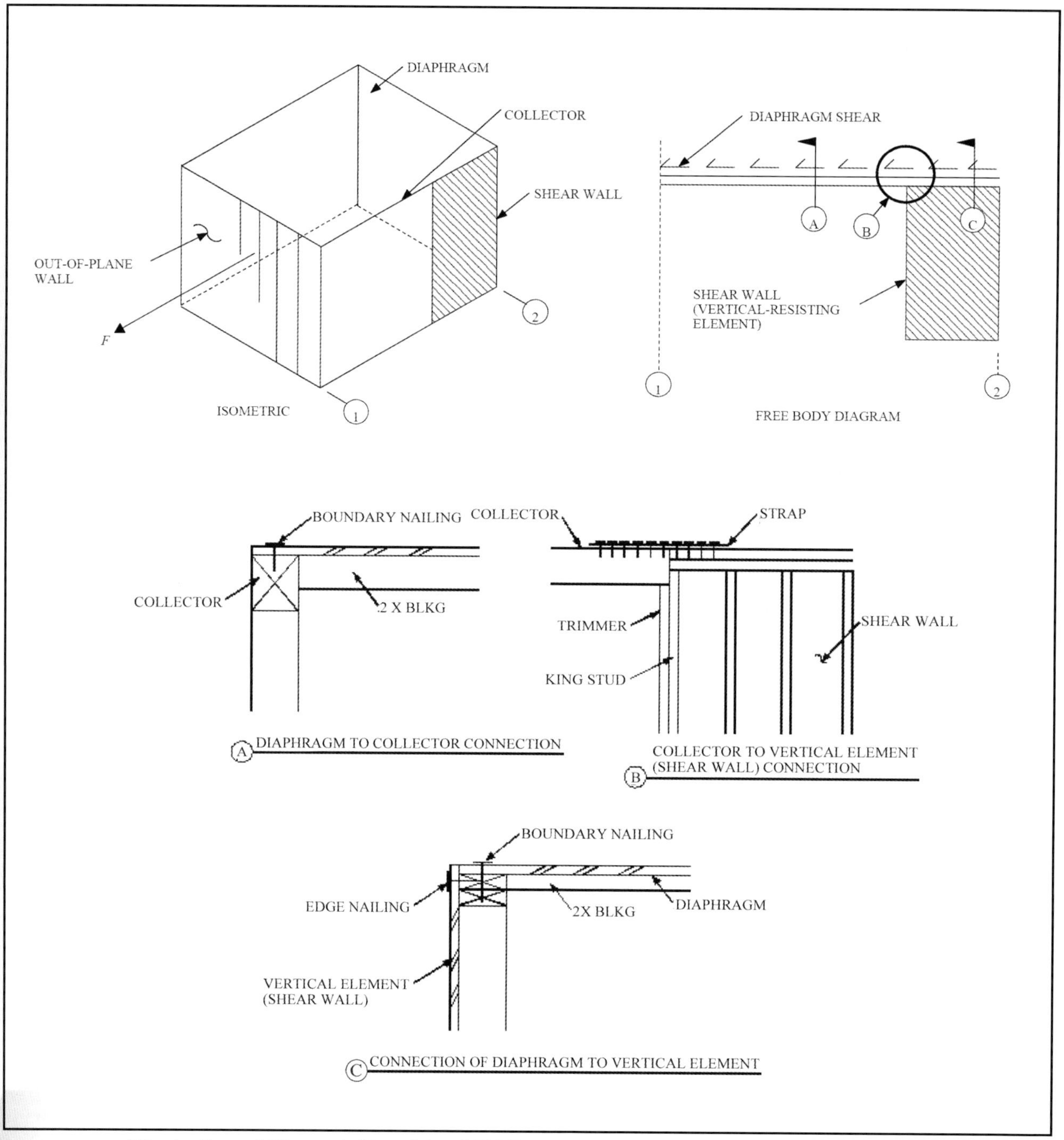

Illustration of Elements Requiring 25% Increase in Section 12.10.1.1 Design Forces
Courtesy: International Code Council

12.3.4.1 Modification

Conditions Where ρ is 1.0

At a Glance

Conditions where value of ρ is 1.0 are expanded to include design of structural walls for out-of-plane bending

2010 Standard

12.3.4.1 Conditions Where Value of ρ is 1.0

The value of ρ is permitted to equal 1.0 for the following:

1. Structures assigned to Seismic Design Category B or C.
2. Drift calculation and P-delta effects.
3. Design of nonstructural components.
4. Design of nonbuilding structures that are not similar to buildings.
5. Design of collector elements, splices, and their connections for which the seismic load effects including overstrength factor of Section 12.4.3 are used.
6. Design of members or connections where the seismic load effects including overstrength factor of Section 12.4.3 are required for design.
7. Diaphragm loads determined using Eq. 12.10-1.
8. Structures with damping systems designed in accordance with Chapter 18.
9. Design of structural walls for out-of-plane forces, including their anchorage.

Analysis and Commentary

The out-of-plane design of walls has been split into two separate sections: Section 12.11 for structural walls and Section 13.3 for nonstructural walls. Whereas defining ρ = 1.0 for nonstructural walls was covered by Item 3 of Section 12.3.4.1, structural walls designed for out-of-plane forces were not included in the list.

Additional confusion existed for concrete slender walls (structural) that are designed for secondary P-delta moments as required by ACI 318 Section 14.8. Item No. 2 of Section 12.3.4.1 seemed to indicate that ρ was 1.0 only because the P-delta effects were included in design. Yet, if the structural wall were not slender and thus there were no P-delta effects, ρ could be greater than 1.0.

The intention of the redundancy factor was to penalize the vertical seismic force- resisting systems, such as the shear walls in-plane, for lack of structural redundancy. The intent was not to penalize wall designs out-of-plane for non-redundant seismic force-resisting systems.

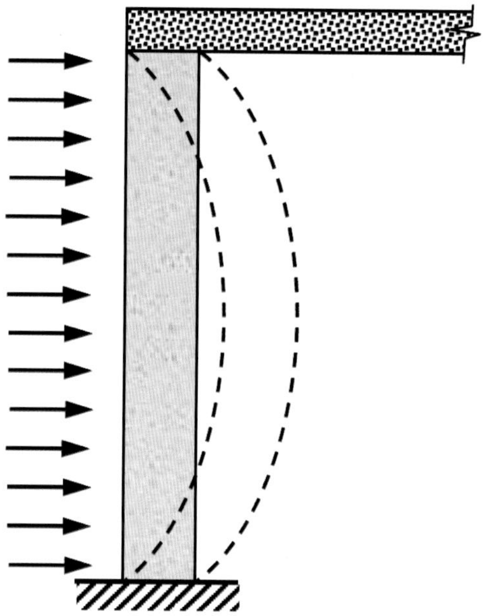

Structural Wall with Out-of-Plane Forces

12.4.3.3 Modification

Allowable Stress Increase for Load Combinations with Overstrength

At a Glance

The provision allowing the duration of load increase to be combined with the allowable stress increase of 1.2 is expanded to include all applicable adjustment factors in accordance with AF&PA NDS.

2010 Standard

12.4.3.3 Allowable Stress Increase for Load Combinations with Overstrength

Where allowable stress design methodologies are used with the seismic load effect defined in Section 12.4.3 applied in load combinations 5, 6, or 8 of Section 2.4.1, allowable stresses are permitted to be determined using an allowable stress increase of 1.2. This increase shall not be combined with increases in allowable stresses or load combination reductions otherwise permitted by this standard or the material reference document except for ~~that combination with the duration of load~~ increases due to adjustment factors ~~permitted~~ in accordance with AF&PA NDS ~~is permitted~~.

Analysis and Commentary

Several adjustment factors for the design of wood construction can result in increases to the reference design values of the AF&PA NDS; some examples are the flat use factor, repetitive member factor, buckling stiffness factor, and bearing area factor. These factors are material-dependent in much the same manner as load duration factor. The prohibition on stress increases when using the allowable stress design load combinations of Section 2.4.1 is intended to prevent increases in stresses already accounted for by the decreases in multiple variable loads incorporated into the load combinations. The adjustment factors listed above, however, are not accounted for because they are not related to design loads.

Table 4.3.1 Applicability of Adjustment Factors for Sawn Lumber

		ASD only	ASD and LRFD										LRFD only		
		Load Duration Factor	Wet Service Factor	Temperature Factor	Beam Stability Factor	Size Factor	Flat Use Factor	Incising Factor	Repetitive Member Factor	Column Stability Factor	Buckling Stiffness Factor	Bearing Area Factor	Format Conversion Factor	Resistance Factor	Time Effect Factor
$F_b' = F_b$	×	C_D	C_M	C_t	C_L	C_F	C_{fu}	C_i	C_r	-	-	-	K_F	ϕ_b	λ
$F_t' = F_t$	×	C_D	C_M	C_t	-	C_F	-	C_i	-	-	-	-	K_F	ϕ_t	λ
$F_v' = F_v$	×	C_D	C_M	C_t	-	-	-	C_i	-	-	-	-	K_F	ϕ_v	λ
$F_{c\perp}' = F_{c\perp}$	×	-	C_M	C_t	-	-	-	C_i	-	-	-	C_b	K_F	ϕ_c	λ
$F_c' = F_c$	×	C_D	C_M	C_t	-	C_F	-	C_i	-	C_p	-	-	K_F	ϕ_c	λ
$E' = E$	×	-	C_M	C_t	-	-	-	C_i	-	-	-	-	-	-	-
$E_{min}' = E_{min}$	×	-	C_M	C_t	-	-	-	C_i	-	-	C_T	-	K_F	ϕ_s	-

AF&PA National Design Specification (NDS) Table 4.3.1

Table 12.6-1

Modification

Permitted Analytical Procedures

At a Glance

Table 12.6-1 is revised to eliminate unnecessary complexity and to use structural height, instead of T_s, as a trigger for requiring dynamic analysis in some cases.

2010 Standard

Table 12.6-1 Permitted Analytical Procedures

Seismic Design Category	Structural Characteristics	Equivalent Lateral Force Analysis, Section 12.8[a]	Modal Response Spectrum Analysis, Section 12.9[a]	Seismic Response History Procedures, Chapter 16[a]
B, C	~~Occupancy Category I or II buildings of light-framed construction not exceeding 3 stories in height~~	~~P~~	~~P~~	~~P~~
	~~Other Occupancy Category I or II buildings not exceeding 2 stories in height~~	~~P~~	~~P~~	~~P~~
	All ~~other~~ structures	P	P	P
D, E, F	~~Occupancy Category I or II buildings of light-framed construction not exceeding 3 stories in height~~	~~P~~	~~P~~	~~P~~
	Risk Category ~~Other Occupancy~~ I or II buildings not exceeding 2 stories ~~in height~~ above the base	P	P	P
	Structures of light frame construction	P	P	P
	~~Regular~~ Structures with no structural irregularities and not exceeding 160 feet in structural height ~~$T < 3.5\,T_s$ and all structures of light frame construction~~	P	P	P
	Structures exceeding 160 feet in structural height with no structural irregularities and with $T < 3.5\,T_s$	P	P	P
	~~Irregular~~ Structures ~~with $T < 3.5\,T_s$~~ not exceeding 160 feet in structural height and having only horizontal irregularities of Type 2, 3, 4, or 5 ~~of~~ in Table 12.3-1 or vertical irregularities of Type 4, 5a, or 5b ~~of~~ in Table 12.3-2	P	P	P
	All other structures	NP	P	P

[a] P: Permitted; NP: Not Permitted; $T_s = S_{D1}/S_{DS}$.

Analysis and Commentary:

There are two significant changes to Table 12.6-1. First the table is revised to eliminate unnecessary complexity. For SDC B and C buildings, the ASCE 7-05 table allowed all analysis procedures all the time. However, three rows in the upper portion of the table were used to communicate this. These three rows have been consolidated into one row. Also, in the first row applicable to SDC D, E, and F, "Occupancy Category I or II buildings of light-framed construction not exceeding 3 stories in height" were exempted from dynamic analysis. This was redundant because the third row applicable to SDC D, E, and F exempted all light-framed buildings. This redundancy is removed in ASCE 7-10.

The second significant change is the introduction of a new threshold for determining whether or not dynamic analysis is required. For the convenience of the user, ASCE 7-10 establishes a new period-independent threshold of 160 feet below which structures without certain irregularities are not required to be subject to dynamic analysis, because higher mode effects are not going to be significant. For regular structures exceeding 160 feet in height, higher mode effects are still judged unlikely to be important as long as the period remains less than the previous threshold of $3.5T_s$ and dynamic analysis is not required. The basis for the $3.5T_s$ limitation is that the higher modes become more dominant in taller buildings (Lopez and Cruz, 1996; Chopra, 2007), and as a result, the ELF method may underestimate the design base shear and may not correctly predict the vertical distribution of seismic forces in taller buildings.

Dynamic analysis continues to be required at times for shorter buildings with certain structural irregularities because those tend to cause undesirable concentrations of inelastic displacements at certain locations. These structural irregularities are: horizontal irregularity type 1a and 1b (torsional and extreme torsional); vertical irregularity type 1a and 1b (soft story and extreme soft story); vertical irregularity type 2 (weight/mass) and vertically irregularity type 3 (geometric). ELF is not allowed for buildings with the listed irregularities because the procedure is based on an assumption of a gradually varying distribution of mass and stiffness along the height and negligible torsional response.

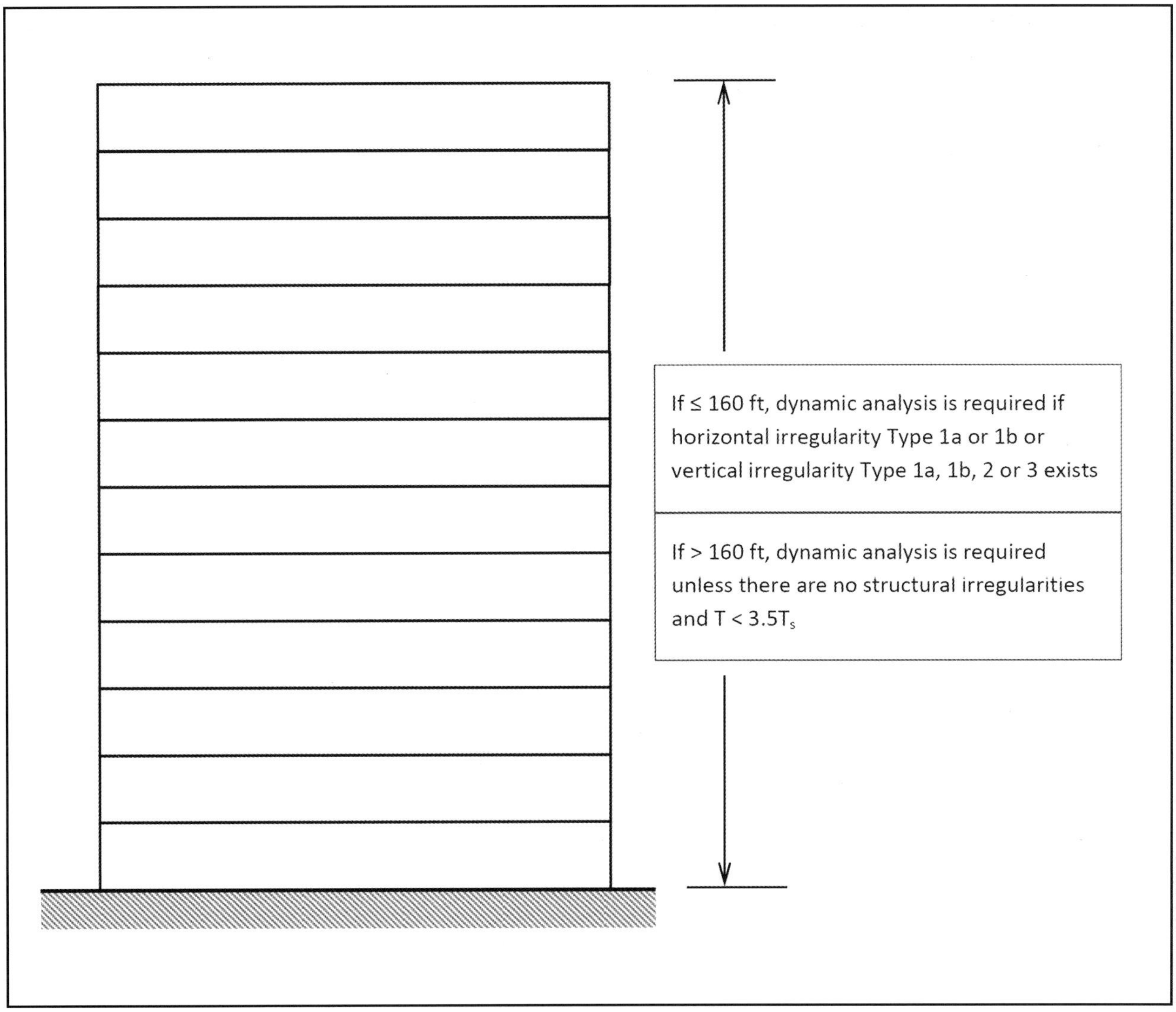

Equivalent Lateral Force (ELF) Analysis or Dynamic Analysis?

12.7.2, 12.14.8.1, 15.3.1, 15.3.2, 15.4.3

Modification and Clarification

Effective Seismic Weight

At a Glance

What is required to be included in the effective seismic weight of a building as well as a nonbuilding structure is better defined.

2010 Standard

12.7.2 Effective Seismic Weight

The effective seismic weight, W, of a structure shall include the ~~total~~ dead load, as defined in Section 3.1, above the base and other loads above the base as listed below:

1. In areas used for storage, a minimum of 25 percent of the floor live load shall be included. ~~(floor live load in public garages and open parking structures need not be included).~~

   EXCEPTIONS:

   a. Where the inclusion of storage loads adds no more than 5% to the effective seismic weight at that level, it need not be included in the effective seismic weight.

   b. Floor live load in public garages and open parking structures need not be included.

2. Where provision for partitions is required by Section 4.2.2 in the floor load design, the actual partition weight or a minimum weight of 10 psf (0.48 kN/m²) of floor area, whichever is greater.
3. Total operating weight of permanent equipment.
4. Where the flat roof snow load, P_f, exceeds 30 psf (1.44 kN/m²), 20 percent of the uniform design snow load, regardless of actual roof slope.
5. Weight of landscaping and other materials at roof gardens and similar areas.

12.14.8.1 Seismic Base Shear

The seismic base shear, V, in a given direction shall be determined in accordance with Eq. 12.14-11:

$$V = \frac{FS_{DS}}{R}W \qquad (12.14\text{-}11)$$

where

$$S_{DS} = \frac{2}{3}F_a S_s$$

where F_a is permitted to be taken as 1.0 for rock sites, 1.4 for soil sites, or determined in accordance with Section 11.4.3. For the purpose of this section, sites are permitted to be considered to be rock if there is no more than 10 ft (3 m) of soil between the rock surface and the bottom of spread footing or mat foundation. In calculating S_{DS}, S_s shall be in accordance with Section 11.4.1, but need not be taken larger than 1.5.

F = 1.0 for buildings that are one story above grade plane

F = 1.1 for buildings that are two stories above grade plane

F = 1.2 for buildings that are three stories above grade plane

R = the response modification factor from Table 12.14-1

W = effective seismic weight of the structure that ~~shall~~ includes the ~~total~~ dead load, as defined in Section 3.1, above grade plane and other loads above grade plane as listed in the following text:

1. In areas used for storage, a minimum of 25 percent of the floor live load shall be included. ~~(floor live load in public garages and open parking structures need not be included).~~

   EXCEPTIONS:

   a. Where the inclusion of storage loads adds no more than 5% to the effective seismic weight at that level, it need not be included in the effective seismic weight.

   b. Floor live load in public garages and open parking structures need not be included.
2. Where provision for partitions is required by Section 4.2.2 in the floor load design, the actual partition weight, or a minimum weight of 10 psf (0.48 kN/m^2) of floor area, whichever is greater.
3. Total operating weight of permanent equipment.
4. Where the flat roof snow load, P_f, exceeds 30 psf (1.44 kN/m^2), 20 percent of the uniform design snow load, regardless of actual roof slope.
5. Weight of landscaping and other materials at roof gardens and similar areas.

15.3.1 Less Than 25 percent Combined Weight Condition

For the condition where the weight of the nonbuilding structure is less than 25 percent of the combined effective seismic weights of the nonbuilding structure and supporting structure, the design seismic forces of the nonbuilding structure shall be determined in accordance with Chapter 13 where the values of R_p and a_p shall be determined in accordance to Section 13.1.5. The supporting structure shall be designed in accordance with the requirements of Chapter 12 or Section 15.5 as appropriate with the weight of the nonbuilding structure considered in the determination of the effective seismic weight, W.

15.3.2 Greater Than or Equal to 25 percent Combined Weight Condition

For the condition where the weight of the nonbuilding structure is equal to or greater than 25 percent of the combined effective seismic weights of the nonbuilding structure and supporting structure, an analysis combining the structural characteristics of both the nonbuilding structure and the supporting structures shall be performed to determine the seismic design forces as follows:

(no changes to remainder of section)

15.4.3 Loads

The seismic effective weight W for nonbuilding structures shall include ~~all~~ the dead load and other loads as defined for structures in Section 12.7.2. For purposes of calculating design seismic forces in nonbuilding structures, W also shall include all normal operating contents for items such as tanks, vessels, bins, hoppers, and the contents of piping. *W* shall include snow and ice loads where these loads constitute 25 percent or more of *W* or where required by the authority having jurisdiction based on local environmental characteristics.

Analysis and Commentary

What is required to be included in the effective seismic weight is better defined. In ASCE 7-05, it was the total dead load of the structure plus certain other loads as detailed in the four items of ASCE 7-05 Section 12.7.2. The "total dead load" of a structure implied that the weight of the structure below the seismic base was also required to be included, which was not the intent. The "total dead load" was also not defined but "dead load" is defined in Section 3.1.1. The effective seismic weight is now defined as the dead load, as defined in Section 3.1, above the base, plus the other loads specified in Items 1 through 5 of Section 12.7.2, also above the base.

In Item 1 of Section 12.7.2, the requirement is that 25 percent of the floor live load in areas used for storage be included in the effective seismic weight. This is limited to storage areas where inclusion has a greater than 5% effect on mass at a level.

A new Item 5 consisting of the weight of landscaping and other materials at roof gardens and similar areas is added because the definition of dead load in Section 3.1.1 does not clearly include the weight of these materials, which will contribute to the seismic mass.

The provisions for effective seismic weight in Section 12.14.8.1 are modified for consistency with Section 12.7.2. Instead of including loads above the base of the structure in the effective seismic weight, however, loads above grade plane are included for consistency with Item 3 of Section 12.14.1.1.

Sections 15.3.1 and 15.3.2 are modified to specify the combined effective seismic weights of the nonbuilding structure and supporting structure. Section 15.4.3 is also modified to recognize that effective seismic weight, as defined in Section 12.7.2, is not limited to dead load.

Weight of landscaping and other materials at roof garden is to be included in effective seismic weight.

12.7.3 Clarification and Addition

Structural Modeling

At a Glance

The applicability of required consideration of cracked section properties in concrete and masonry structures and panel zone deformations in steel moment frames is clarified. A new exception exempts structures with Type 4 horizontal structural irregularity having flexible diaphragms from 3-D analysis requirement.

2010 Standard

12.7.3 Structural Modeling

A mathematical model of the structure shall be constructed for the purpose of determining member forces and structure displacements resulting from applied loads and any imposed displacements or P-Delta effects. The model shall include the stiffness and strength of elements that are significant to the distribution of forces and deformations in the structure and represent the spatial distribution of mass and stiffness throughout the structure.

~~Structures that have horizontal structural irregularity Type 1a, 1b, 4, or 5 of Table 12.3-1 shall be analyzed using a 3-D representation. Where a 3-D model is used, a minimum of three dynamic degrees of freedom consisting of translation in two orthogonal plan directions and torsional rotation about the vertical axis shall be included at each level of the structure. Where the diaphragms have not been classified as rigid or flexible in accordance with Section 12.3.1, the model shall include representation of the diaphragm's stiffness characteristics and such additional dynamic degrees of freedom as are required to account for the participation of the diaphragm in the structure's dynamic response.~~ In addition, the model shall comply with the following:

a. Stiffness properties of concrete and masonry elements shall consider the effects of cracked sections.

b. For steel moment frame systems, the contribution of panel zone deformations to overall story drift shall be included.

<u>Structures that have horizontal structural irregularity Type 1a, 1b, 4, or 5 of Table 12.3-1 shall be analyzed using a 3-D representation. Where a 3-D model is used, a minimum of three dynamic degrees of freedom consisting of translation in two orthogonal plan directions and rotation about the vertical axis shall be included at each level of the structure. Where the diaphragms have not been classified as rigid or flexible in accordance with Section 12.3.1, the model shall include representation of the diaphragm's stiffness characteristics and such additional dynamic degrees of freedom as are required to account for the participation of the diaphragm in the structure's dynamic response.</u>

<u>**EXCEPTION:** Analysis using a 3-D representation is not required for structures with flexible diaphragms that have Type 4 horizontal structural irregularities.</u>

Analysis and Commentary

The ASCE 7-05 text implied that the required consideration of cracked section properties in concrete and masonry structures and panel zone deformation in steel moment frames was limited to structures with one of the horizontal structural irregularities listed in the second paragraph. This was due to the placement of the requirement at the end of the second paragraph, which was limited in scope to structures with one or more of these irregularities. The proposal relocates the requirement to immediately follow the end of the first paragraph so that it applies to all structures, which is the intent.

The addition of the exception is a more substantive change and is explained below.

Three-dimensional structural modeling is warranted for structures with horizontal structural irregularities that will have load effects on vertical seismic force-resisting elements in a direction other than parallel to the direction under consideration (orthogonal effects). This is typically the case for structures with rigid diaphragms. For structures with flexible diaphragms, however, such effects are not likely to occur in most cases to an extent that warrants such modeling.

For horizontal irregularities Type 1a and 1b (torsional and extreme torsional), the issue is moot since torsion need not be considered in structures with flexible diaphragms. Sections 12.8.4.1 and 12.8.4.2 on inherent torsion and accidental torsion, respectively, exempt flexible diaphragms from consideration. Table 12.3-1 on horizontal structural irregularities limits application of horizontal irregularities Type 1a and 1b to structures in which the diaphragms are rigid or semirigid. For horizontal irregularity Type 5 (nonparallel seismic force-resisting systems), substantial orthogonal effects are possible regardless of the type of diaphragm, since the vertical seismic force-resisting elements are not parallel with the major orthogonal axes of the seismic force-resisting system. For horizontal irregularity Type 4 (out-of-plane offsets) in a structure with rigid or semirigid diaphragms, substantial orthogonal effects are also possible due to the eccentricity in the vertical load path of the seismic force-resisting system caused by the horizontal offsets of the vertical elements from story to story.

For horizontal irregularity Type 4 (out-of-plane offsets) in a structure with flexible diaphragms, however, substantial orthogonal effects are not likely. The eccentricity in the vertical load path will cause a redistribution of seismic design forces from the vertical elements at the story above to the vertical elements essentially in the same direction at the story below. The effect on the vertical elements in the orthogonal direction will be minimal.

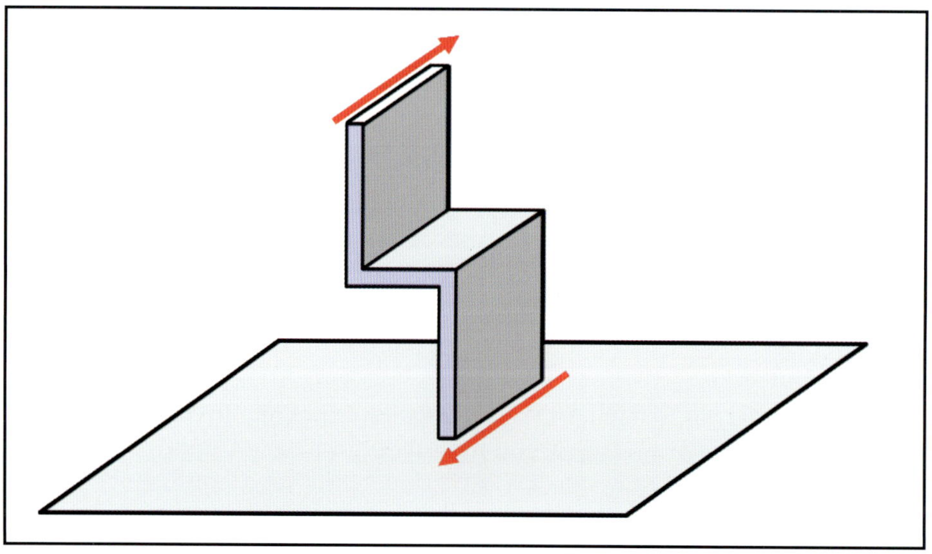

Structures with out-of-plane offsets of vertical lateral force-resisting elements (horizontal structural irregularity Type 4) are exempt from 3-D analysisrequirement if diaphragms are flexible.

Courtesy: Federal Emergency Management Agency

12.8.1.1

Addition (Reinstatement)

Minimum Design Base Shear

At a Glance

The minimum design base shear of $0.044 S_{DS} I_e W$, applicable for Seismic Design Categories B through F, is reinstated from ASCE 7-02 in the equivalent lateral force procedure.

2010 Standard

12.8.1.1 Calculation of Seismic Response Coefficient

The seismic response coefficient, C_s, shall be determined in accordance with Eq. 12.8-2.

$$C_s = \frac{S_{DS}}{\left(\frac{R}{I_e}\right)} \qquad \text{(Eq. 12.8-2)}$$

where

S_{DS} = the design spectral response acceleration parameter in the short period range as determined from Section 11.4.4 or 11.4.7

R = the response modification factor in Table 12.2-1

I_e = the importance factor determined in accordance with Section 11.5.1

The value of C_s computed in accordance with Eq. 12.8-2 need not exceed the following:

$$C_s = \frac{S_{D1}}{T\left(\frac{R}{I_e}\right)} \quad \text{for } T \leq T_L \qquad \text{(Eq. 12.8-3)}$$

$$C_s = \frac{S_{D1} T_L}{T^2\left(\frac{R}{I_e}\right)} \quad \text{for } T > T_L \qquad \text{(Eq. 12.8-4)}$$

C_s shall not be less than

$C_s =$ ~~0.01~~ $\underline{0.044 S_{DS} I_e \geq 0.01}$ (Eq. 12.8-5)

In addition, for structures located where S_1 is equal to or greater than 0.6g, C_s shall not be less than

$$C_S = 0.5S_1/(R/I_e) \qquad \text{(Eq. 12.8-6)}$$

where I_e and R are as defined in Section 12.8.1.1 and

S_{D1} = the design spectral response acceleration parameter at a period of 1.0 s, as determined from Section 11.4.4 or 11.4.7

T = the fundamental period of the structure (s) determined in Section 12.8.2

T_L = long-period transition period (s) determined in Section 11.4.5

S_1 = the mapped maximum considered earthquake spectral response acceleration parameter determined in accordance with Section 11.4.1 or 11.4.7

Analysis and Commentary

The minimum design base shear of $0.044S_{DS}I_eW$, applicable for Seismic Design Categories (SDC) B through F, was part of ASCE 7-02 and the 2000 and 2003 IBC. However, when the third (constant-displacement) branch, starting at T_L, was added to ASCE 7-05, this minimum base shear was deleted in favor of just 1% of weight, which is a structural integrity minimum, applicable irrespective of SDC. The basis was that the long-period structure was now being directly addressed by the constant-displacement branch of the design spectrum, so that there was no need for an arbitrary minimum value.

In the course of the ATC-63 project (*Quantification of Building System Performance and Response Parameters*), a large number of ordinary as well as special moment frames of concrete were analyzed by state-of-the-art dynamic analysis procedures – each frame under a large number of pairs of earthquake ground motions. The analyses disturbingly showed story mechanisms forming even in the special moment frames, which satisfied the strong column – weak beam requirement, early into earthquake excitations. After considerable deliberations, the suggestion emerged that these frames, designed by ASCE 7-05, be designed instead by ASCE 7-02, in effect reinstating the minimum design base shear requirement of $0.044S_{DS}I_eW$. When this was done, the problem went away, leading to the inescapable conclusion that the removal of this minimum base shear had been a mistake. ASCE processed a Supplement No. 2 to ASCE 7-05, reinstating this minimum design base shear. Supplement No. 2 has been adopted by the 2009 IBC. ASCE 7-10 has now incorporated Supplement No. 2 in its body.

The minimum design base shear of $0.044S_{DS}I_e$ is also added to Equations 15.4-1 and 15.4-2 for nonbuilding structures not similar to buildings. Because nonbuilding structures not similar to buildings have low R-values compared to the special reinforced concrete moment frames studied in ATC-63, the ASCE 7 standards committee chose not to restore the high minimum base shears for nonbuilding structures not similar to buildings found in ASCE 7-02. In many cases, these previous minimum base shears gave many nonbuilding structures not similar to buildings effective R-values less than 1.0. Therefore, the committee believed that the minimum base shear equation of $0.044S_{DS}I_e$ used for buildings should also be applied to nonbuilding structures not similar to buildings.

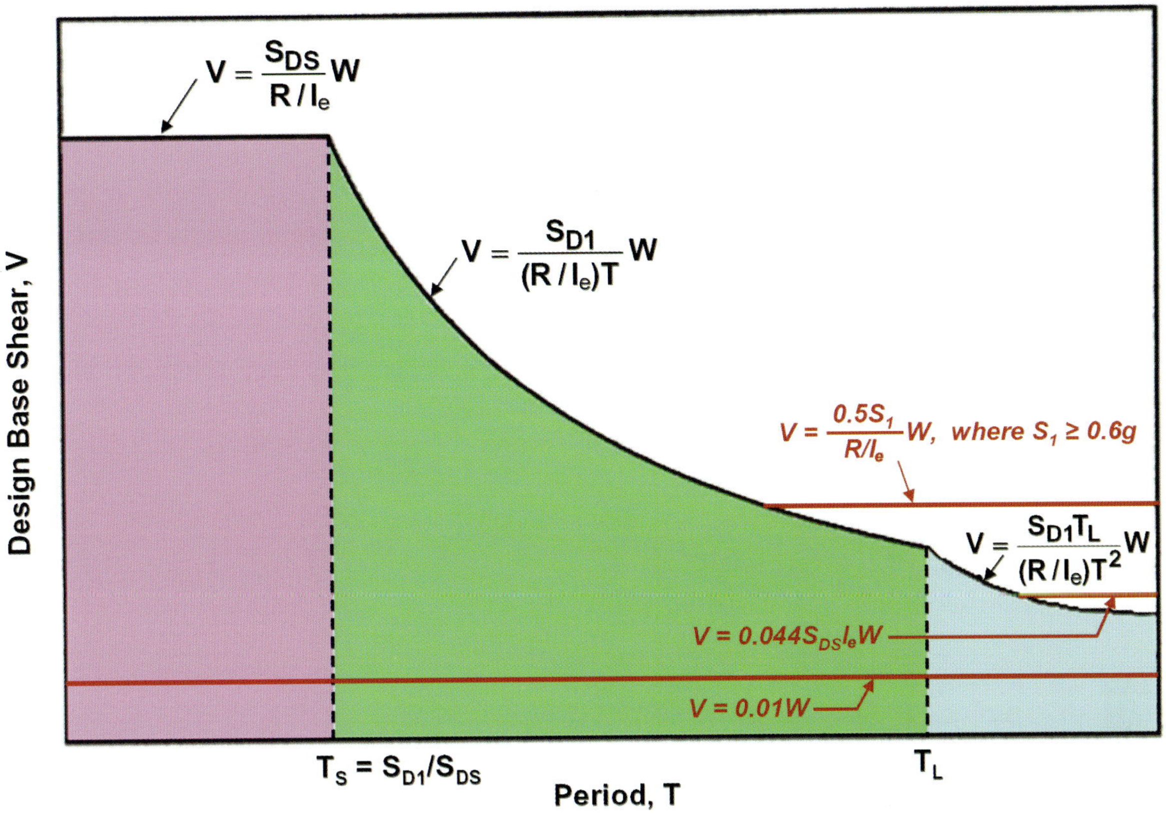

ASCE 7-10 Design Response Spectrum for Equivalent Lateral Force Procedure

Courtesy: S.K. Ghosh Associates Inc.

12.8.1.1 Clarification

S_{DS}, S_{D1} in Seismic Response Coefficient Computation

At a Glance

Clarification is provided that S_{DS} and S_{D1} for computation of seismic design base shear can be obtained from site-specific ground motion procedures.

2010 Standard

12.8.1.1 Calculation of Seismic Response Coefficient

The seismic response coefficient, C_s, shall be determined in accordance with Eq. 12.8-2.

$$C_s = \frac{S_{DS}}{\left(\frac{R}{I_e}\right)} \quad (12.8\text{-}2)$$

where

S_{DS} = the design spectral response acceleration parameter in the short period range as determined from Section 11.4.4 or 11.4.7

R = the response modification factor in Table 12.2-1

I_e = the importance factor determined in accordance with Section 11.5.1

The value of C_s computed in accordance with Eq. 12.8-2 need not exceed the following:

$$C_s = \frac{S_{D1}}{T\left(\frac{R}{I_e}\right)} \quad \text{for } T \le T_L \quad (12.8\text{-}3)$$

$$C_s = \frac{S_{D1}T_L}{T^2\left(\frac{R}{I_e}\right)} \quad \text{for } T > T_L \quad (12.8\text{-}4)$$

C_s shall not be less than

$$C_s = 0.044 S_{DS} I_e \ge 0.01 \quad (12.8\text{-}5)$$

In addition, for structures located where S_1 is equal to or greater than 0.6*g*, C_s shall not be less than

$$C_S = 0.5S_1/(R/I_e) \qquad (12.8\text{-}6)$$

Where I_e and R are as defined in Section 12.8.1.1 and

S_{D1} = the design spectral response acceleration parameter at a period of 1.0 s, as determined from Section 11.4.4 or 11.4.7

T = the fundamental period of the structure (s) determined in Section 12.8.2

T_L = long-period transition period (s) determined in Section 11.4.5

S_1 = the mapped maximum considered earthquake spectral response acceleration parameter determined in accordance with Section 11.4.1 or 11.4.7

Analysis and Commentary

In the event that a site-specific response spectrum is used to characterize the design ground motion, S_{DS} and S_{D1} are determined in accordance with Section 11.4.7 rather than 11.4.1. Although it may already be implied that the values obtained using the site-specific procedure (when applicable) may be used, the ASCE 7-10 text makes this more explicit.

The potential for confusion exists particularly if the site-specific response spectrum is used to perform a dynamic analysis using the modal response spectrum procedure of Section 12.9. In this case, S_{DS} and S_{D1} may not be computed when the response spectrum is developed because they are not needed to perform the analysis. However, in accordance with Section 12.9.4, the designer must compare the base shear obtained from the dynamic analysis with that given by the equivalent lateral force procedure and scale the results if necessary. As currently written, the code leads the user back to the general ground acceleration parameters of Section 11.4.4 for computation of the base shear from the equivalent static procedure rather than the site-specific values.

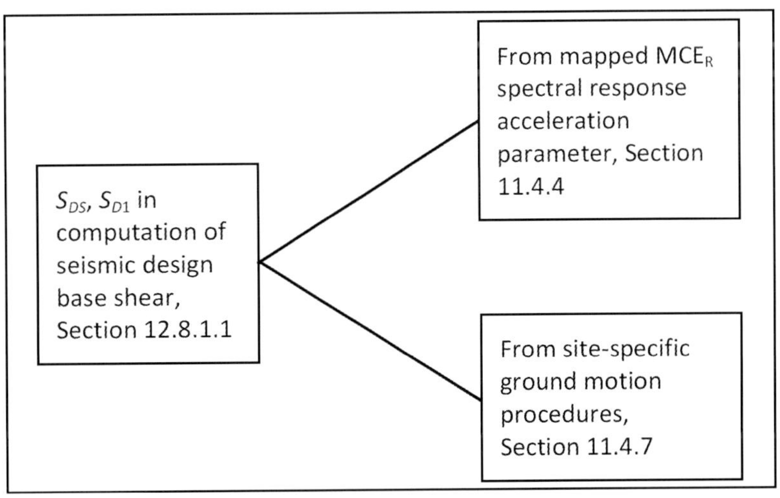

Either Section 11.4.4 or 11.4.7 may be used to determine the seismic design base shear when using the equivalent lateral force procedure of Section 12.8.

12.8.2.1, Table 12.8-2

Clarification and Addition

Approximate Fundamental Period

At a Glance

Determination of approximate fundamental period of steel eccentrically braced frames is clarified and steel buckling-restrained braced frames are added to Table 12.8-2.

2010 Standard

12.8.2.1 Approximate Fundamental Period

The approximate fundamental period (T_a), in s, shall be determined from the following equation:

$$T_a = C_t h_n^x \quad \text{(Eq. 12.8-7)}$$

where h_n is the structural height as defined in Section 11.2 and the coefficients C_t and x are determined from Table 12.8-2

Table 12.8- 2 Values of Approximate Period Parameters C_t and x

Structure Type	C_t	x
Moment resisting frame systems in which the frames resist 100% of the required seismic force and are not enclosed or adjoined by components that are more rigid and will prevent the frames from deflecting where subjected to seismic forces:		
Steel Moment Resisting Frames	0.028 (0.0724)[a]	0.8
Concrete Moment Resisting Frames	0.016 (0.0466)[a]	0.9
Steel e~~E~~ccentrically braced ~~steel~~ frames in accordance with Table 12.2-1 lines B1 or D1	0.03 (0.0731)[a]	0.75
Steel buckling-restrained braced frames	0.03 (0.0731)[a]	0.75
All other structural systems	0.02 (0.0488)[a]	0.75

[a] Metric equivalents are shown in parentheses.

Analysis and Commentary

The longer predicted periods represented by the coefficient $C_t = 0.03$ for steel eccentrically braced frames (EBFs) are appropriate where significant eccentricities exist such as is the case when designing eccentrically braced frames in accordance with AISC 341. Without the added language in Table 12.8-2, minimal eccentricities could be used to justify the longer period, which would unconservatively result in a reduced base shear. The added language provides clarification and ensures that significant EBF eccentricities exist, which are consistent with longer periods.

The steel buckling restrained braced frame (BRBF) system was first approved for the 2003 NEHRP *Provisions*. The values for the approximate period parameters C_t and x were also approved as part of that original BSSC proposal. It appears to have been an oversight that these parameters were not carried forward into ASCE 7-05. Currently, these two factors can be found in Appendix R of AISC 341-05. These will be removed in the next edition of AISC 341 in view of this change in ASCE 7-10.

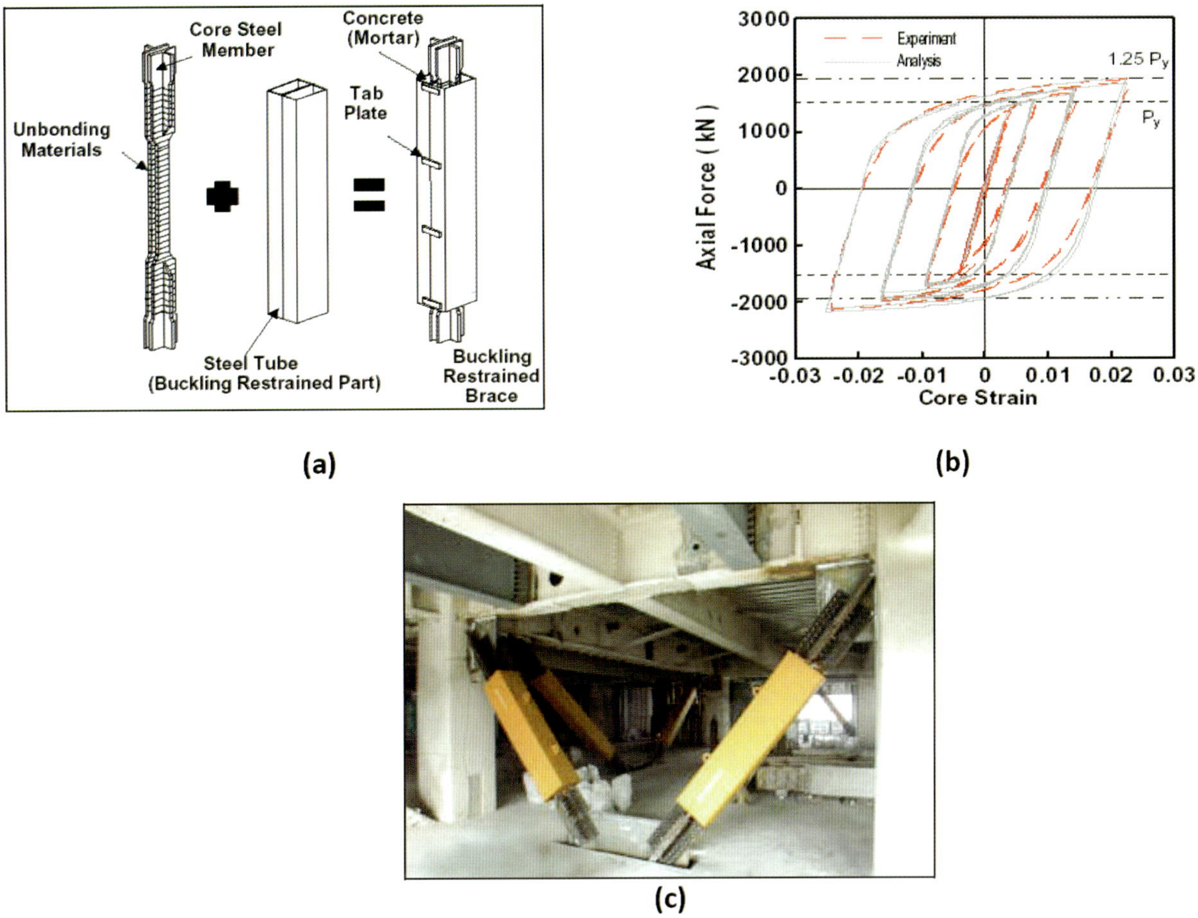

(a) Typical Configuration of Buckling Restrained Brace (BRB), (b) Typical Load-Displacement Behavior of a BRB, and (c) BRBs in Taipei County Administration Building, Taiwan
(reprinted with permission from Dr. K. C. Tsai, National Taiwan University, Taipei)

12.8.2.1 Modification

Approximate Period Equation Based on Number of Stories

At a Glance

The word "average" is added to story height so that the approximate period equation based on number of stories can be used if the average story height is at least 10 feet.

2010 Standard

12.8.2.1 Approximate Fundamental Period

The approximate fundamental period (T_a), in s, shall be determined from the following equation:

$$T_a = C_t h_n^x \qquad (12.8\text{-}7)$$

where h_n is the structural height as defined in Section 11.2 and the coefficients C_t and x are determined from Table 12.8-2.

Alternatively, it is permitted to determine the approximate fundamental period (T_a), in s, from the following equation for structures not exceeding 12 stories above the base as defined in Section 11.2 where the seismic force-resisting system consists entirely of concrete or steel moment resisting frames and the average story height~~, h_{sx},~~ is at least 10 ft (3 m):

$$T_a = 0.1\,N \qquad (12.8\text{-}8)$$

where N = number of stories above the base.

(No change in remainder of section)

Analysis and Commentary

In defining the applicability of Eq. (12.8-8), the 10-ft minimum story height is revised such that it is an average story height.

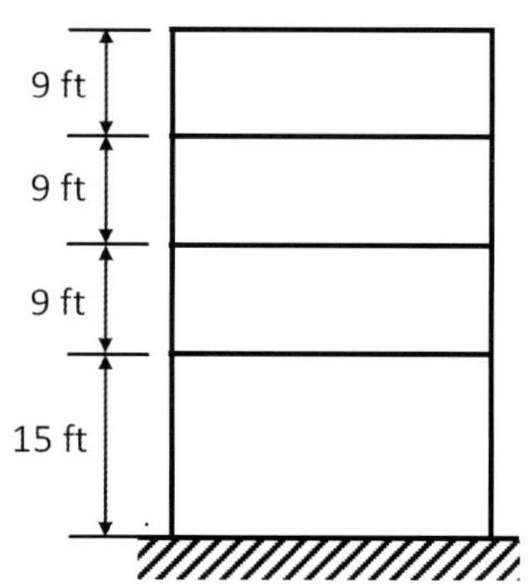

For illustration,
Minimum Story Height = 9 ft
Average Story Height = 10.5 ft
ASCE 7-05 Equation 12.8-8 does not apply
ASCE 7-10 Equation 12.8-8 applies

12.8.4.3

Modification and Deletion

Amplification of Accidental Torsion

At a Glance

The exemption for structures of light frame construction from the amplification of accidental torsion requirement is discontinued.

2010 Standard

12.8.4.3 Amplification of Accidental Torsional Moment

Structures assigned to Seismic Design Category C, D, E, or F, where Type 1a or 1b torsional irregularity exists as defined in Table 12.3-1 shall have the effects accounted for by multiplying M_{ta} at each level by a torsional amplification factor (A_x) as illustrated in Fig. 12.8-1 and determined from the following equation:

$$A_x = (\delta_{max} / 1.2\delta_{avg})^2 \qquad (12.8\text{-}14)$$

where

δ_{max} = the maximum displacement at Level x computed assuming $A_x = 1$ (in. or mm)

δ_{avg} = the average of the displacements at the extreme points of the structure at Level x computed assuming $A_x = 1$ (in. or mm)

~~EXCEPTION: The accidental torsional moment need not be amplified for structures of light-frame construction.~~

The torsional amplification factor (A_x) shall not be less than 1 and is not required to exceed 3.0. The more severe loading for each element shall be considered for design.

Analysis and Commentary

Where wood-frame diaphragms are designed as rigid diaphragms (one example is diaphragms in open-front structures), amplification of torsion should apply. Commentary to AF&PA's *Special Design Provisions for Wind and Seismic* indicates application of provisions of ASCE 7 for increased forces due to presence of irregularities, in addition to special prescriptive limits on diaphragm aspect ratio for open front structures.

Because it is possible for A_x to be less than 1 per Eq. 12.8-14, text is added to clearly indicate that A_x shall not be less than 1.

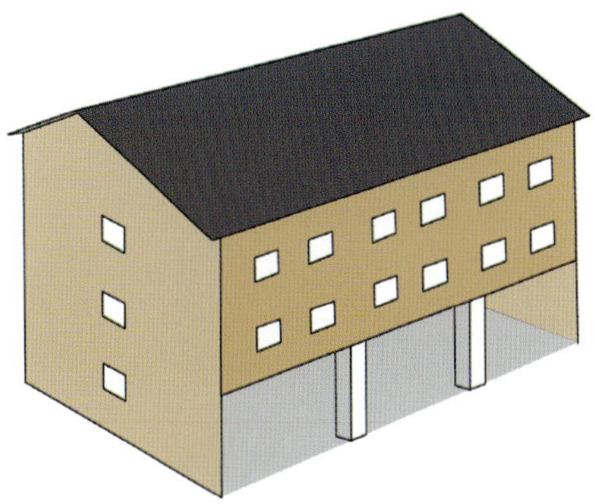

Light-frame Construction with Torsional Irregularity

12.8.6, 12.12.1 Modification

Story Drift Determination

At a Glance

Aspects of story drift determination are clarified.

2010 Standard

12.8.6 Story Drift Determination

The design story drift (Δ) shall be computed as the difference of the deflections at the centers of mass at the top and bottom of the story under consideration. See Fig. 12.8-2. Where centers of mass do not align vertically, it is permitted to compute the deflection at the bottom of the story based on the vertical projection of the center of mass at the top of the story. Where allowable stress design is used, Δ shall be computed using the strength level seismic forces specified in Section 12.8 without reduction for allowable stress design.

For structures assigned to Seismic Design Category C, D, E, or F having horizontal irregularity Type 1a or 1b of Table 12.3-1, the design story drift, Δ, shall be computed as the largest difference of the deflections of vertically aligned points at the top and bottom of the story under consideration along any of the edges of the structure.

The deflections at ~~of~~ Level x (δ_x) (in. or mm) used to compute the design story drift, Δ, ~~at the center of the mass (δ_x) (in. or mm)~~ shall be determined in accordance with the following equation:

$$\delta_x = \frac{C_d \delta_{xe}}{I_e} \qquad (12.8\text{-}15)$$

where

C_d = the deflection amplification factor in Table 12.2-1

δ_{xe} = the deflection~~s~~ at the location required in this section determined by an elastic analysis

I_e = the importance factor determined in accordance with Section 11.5.1

12.12.1 Story Drift Limit

The design story drift (Δ) as determined in Sections 12.8.6, 12.9.2, or 16.1, shall not exceed the allowable story drift (Δ_a) as obtained from Table 12.12-1 for any story. ~~For structures with significant torsional deflections, the maximum drift shall include torsional effects. For structures assigned to Seismic Design Category C, D, E, or F having horizontal irregularity Types 1a or 1b of Table 12.3-1, the design story drift, Δ, shall be computed as the largest difference of the deflections along any of the edges of the structure at the top and bottom of the story under consideration.~~

Analysis and Commentary

Section 12.12.1 focuses on story drift limits and not computation of the story drift demand, Δ. Determination of story drift demand is outlined in Section 12.8.6. Therefore, to consolidate and provide distinct separation between limit and demand, the last sentence in Section 12.12.1 that discusses determination of story drift when horizontal irregularity type 1a or 1b is present is moved to Section

12.8.6. Also, Section 12.8.7 (P-Delta Effects) references Section 12.8.6 and not also Section 12.12.1. The intent is not to limit Δ by taking displacement at the centers of mass for P-delta computation when horizontal irregularity type 1a or 1b is present.

Many computer programs can explicitly provide drift ratios; however, there are many cases and times when the computer programs do not use the same vertically aligned points to compute these ratios, thus yielding an inaccurate measure of drift. A sentence was added in the first paragraph of Section 12.8.6 to permit vertical projections of points when centers of mass do not align vertically. Vertically aligned points are also called for in the second paragraph, which applies to structures assigned to SDC C and above and having torsional or extreme torsional irregularities. In these cases, torsion must be included in deflection computation, so that drifts are based on diaphragm-edge deflections, rather than deflections at the centers of mass.

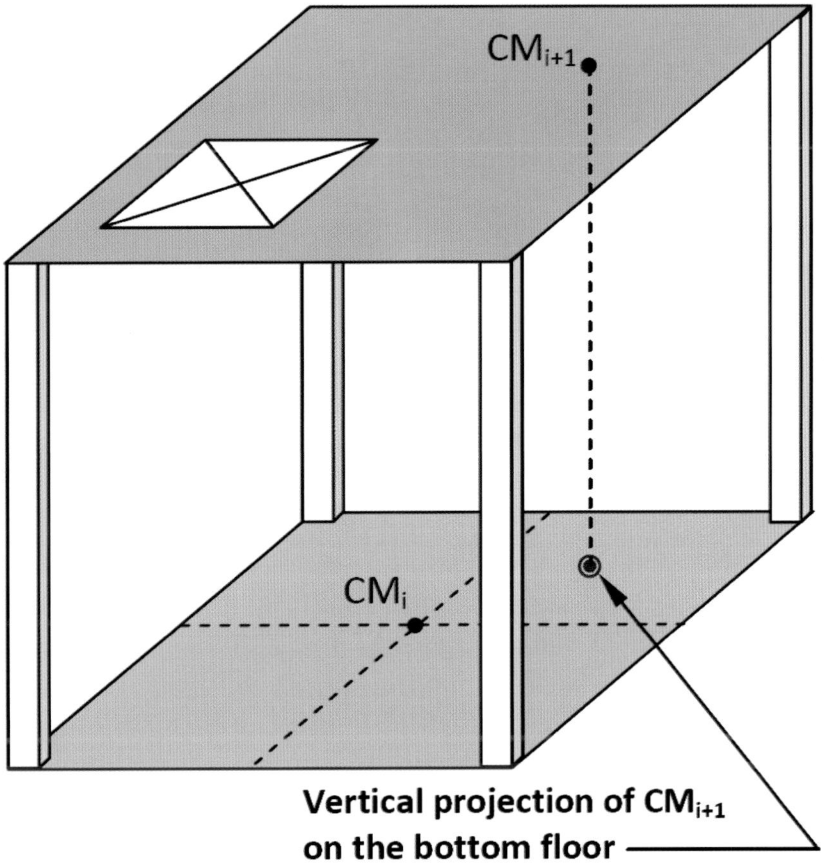

Interstory drift in a building without torsional irregularity where the center of mass of a floor does not align vertically with the center of mass of the floor below.

12.8.6.1 Modification

Minimum Base Shear for Computing Drift

At a Glance

The minimum base shear of $0.044S_{DS}I_eW \geq 0.01W$ is exempted from drift computation.

2010 Standard

12.8.6.1 Minimum Base Shear for Computing Drift.

The elastic analysis of the seismic force-resisting system for computing drift shall be made using the prescribed seismic design forces of Section 12.8.

EXCEPTION: Eq. 12.8-5 need not be considered for computing drift.

Analysis and Commentary

The 1997 *Uniform Building Code* (UBC) exempted the corresponding minimum base shear of $0.11C_aIW$ from drift computation. This was not adopted by ASCE 7-02, ASCE 7-05, or the first four editions of the IBC. Now the exemption is being brought back. This change is significant when it comes to the design of tall buildings.

Tall buildings are typically drift-controlled, rather than strength-controlled. The design of many tall buildings, irrespective of seismic design category, is likely to be governed, in the absence of this exemption, by drift computed under the minimum design base shear given by Eq. (12.8-5). This is considered unjustified, since this minimum design base shear is essentially a minimum strength requirement. The near-fault minimum, as given by Eq. (12.8-6) has a physical basis and is not exempt.

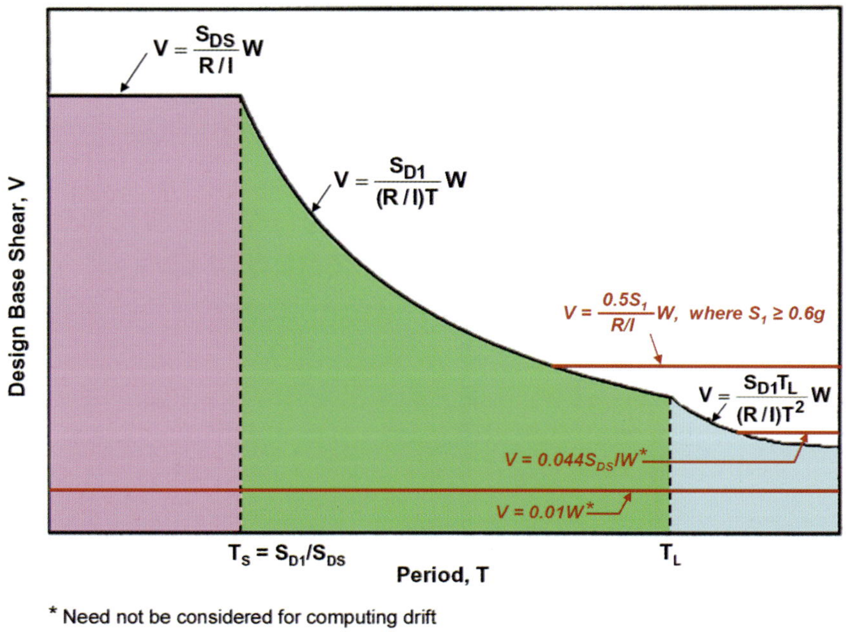

Design Response Spectrum Illustrating Equivalent Lateral Force Procedure Base Shear Equations
Courtesy: S.K. Ghosh Associates Inc.

12.8.7 Modification

P-Delta Effects

At a Glance

The Importance Factor, I_e, is now included in the equation for the stability coefficient, θ.

2010 Standard

12.8.7 P-Delta Effects

P-delta effects on story shears and moments, the resulting member forces and moments, and the story drifts induced by these effects are not required to be considered where the stability coefficient (θ) as determined by the following equation is equal to or less than 0.10:

$$\theta = \frac{P_x \Delta \underline{I_e}}{V_x h_{sx} C_d} \quad (12.8\text{-}16)$$

where

P_x = the total vertical design load at and above Level x (kip or kN); where computing P_x, no individual load factor need exceed 1.0

Δ = the design story drift as defined in Section 12.8.6 occurring simultaneously with V_x (in. or mm)

$\underline{I_e}$ = the importance factor determined in accordance with Section 11.5.1

V_x = the seismic shear force acting between Levels x and $x - 1$ (kip or kN)

h_{sx} = the story height below Level x (in. or mm)

C_d = the deflection amplification factor in Table 12.2-1

(no change in remainder of Section 12.8.7)

Analysis and Commentary

In the opinion of the Seismic Subcommittee of ASCE 7, the Importance Factor, I_e, had been dropped inadvertently from Equation 12.8-16 while transcribing it from the 2003 NEHRP *Provisions* Equation 5.2-16. In the 2003 NEHRP *Provisions*, the importance factor is included in the stability coefficient.

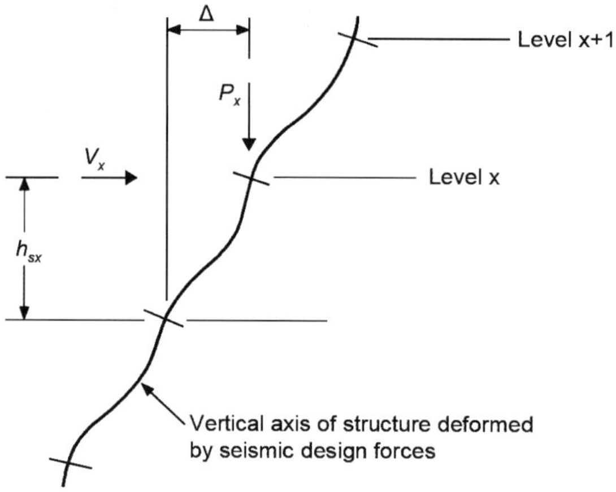

Stability coefficient θ is given by

$$\theta = \frac{P_x \Delta I_e}{V_x h_{sx} C_d}$$

where

$$P_x = \sum_{i=x}^{n} (D+L)_i$$

P-Delta Symbols and Notation

12.9.3

Combined Response Parameters

Modification and Clarification

At a Glance

The combination of response parameters in modal response spectrum analysis is clarified and modified.

2010 Standard

12.9.3 Combined Response Parameters.

The value for each parameter of interest calculated for the various modes shall be combined using ~~either~~ the square root of the sum of the squares (SRSS) method, ~~or~~ the complete quadratic combination (CQC) method, the complete quadratic combination method ~~(CQC)~~ as modified by ~~in accordance with~~ ASCE 4 (CQC-4), or an approved equivalent approach. The CQC or the CQC-4 method shall be used for each of the modal values ~~or~~ where closely spaced modes ~~that~~ have significant cross-correlation of translational and torsional response.

Analysis and Commentary

It is now pointed out to the user of ASCE 7 that the CQC modal response combination method, as it is presented in ASCE 4-98, *Seismic Analysis of Safety-Related Nuclear Structures*, varies slightly from the classical method as formulated by Der Kiureghian and implemented by various commercially available structural analysis software packages.

This difference primarily affects high frequency structures that have significant components of response that are in phase with the ground motion. In these circumstances, using the classical CQC method or the SRSS method for combining maximum modal responses can be unconservative. Furthermore, the commentary to ASCE 4-98 mentions several practical alternate approaches that may be used to achieve the goal of correlating responses that are in phase with the ground motion.

The U.S. Nuclear Regulatory Commission publication: *Reevaluation of Regulatory Guidance on Modal Response Combination Methods for Seismic Response Spectrum Analysis* (NUREG/CR-6645, Brookhaven National Laboratory, 1999) provides a general overview of the various modal response combination methods that have found their way into use. Comparisons have been done between response spectrum and time history analyses using compatible input motions. These studies reveal that modal response combination procedures based on correlation of modes that are in phase with the input motion tend to match more closely the results of the time history analyses than the CQC method by itself.

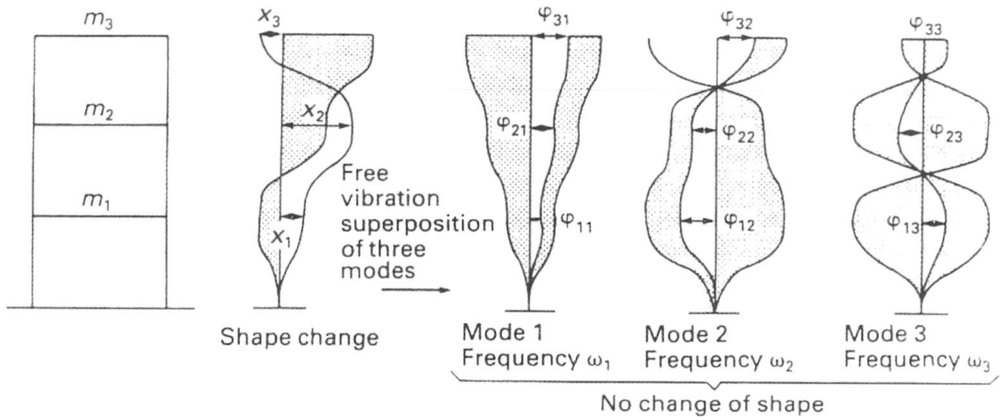

→ SRSS METHOD
→ CQC METHOD
→ CQC-4 METHOD
→ APPROVED EQUIVALENT APPROACH

Methods Permitted to be Used to Combine Modal Response Parameters

12.9.4, 12.9.4.1, 12.9.4.2

Addition

Scaling of Drifts

At a Glance

Provision is added for scaling of drifts where the dynamic base shear is less than the minimum base shear given by Eq. (12.8-6).

2010 Standard

12.9.4 Scaling Design Values of Combined Response

A base shear (V) shall be calculated in each of the two orthogonal horizontal directions using the calculated fundamental period of the structure T in each direction and the procedures of Section 12.8~~,~~. ~~except~~

12.9.4.1 Scaling of Forces

W~~w~~here the calculated fundamental period exceeds $C_u T_a$ in a given direction, ~~then~~ $C_u T_a$ shall be used in lieu of T in that direction. Where the combined response for the modal base shear (V_t) is less than 85 percent of the calculated base shear (V) using the equivalent lateral force procedure, the forces~~, but not the drifts,~~ shall be multiplied by $0.85\frac{V}{V_t}$:

where

V = the equivalent lateral force procedure base shear, calculated in accordance with this section and Section 12.8~~, and~~

V_t = the base shear from the required modal combination

12.9.4.2 Scaling of Drifts

Where the combined response for the modal base shear (V_t) is less than $0.85C_sW$, and where C_s is determined in accordance with Eq. 12.8-6, drifts shall be multiplied by $0.85\frac{C_sW}{V_t}$.

Analysis and Commentary

Provision is added for scaling of drifts where the near-fault minimum base shear equation (Eq. 12.8-6) governs. There was general agreement within the Seismic Subcommittee of ASCE 7 that this was the intent all along and that this requirement is important for long-period structures (for which this minimum base shear equation controls).

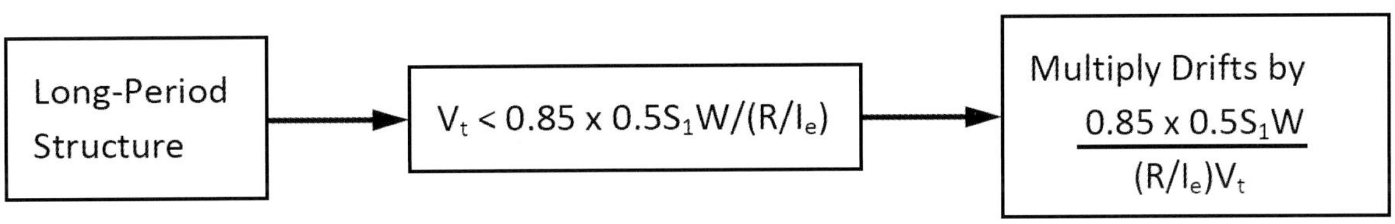

12.4.3.1, 12.10.1.1, 12.10.2.1 Modification and Clarification

Diaphragm and Collector Design Forces

At a Glance

The applicability of the overstrength factor in the determination of collector design forces is clarified. It is also clarified that diaphragm design forces are horizontal earthquake effects, Q_E.

2010 Standard

12.4.3.1 Horizontal Seismic Load Effect with Overstrength Factor

The horizontal seismic load effect with overstrength factor, E_{mh}, shall be determined in accordance with Eq. 12.4-7 as follows:

$$E_{mh} = \Omega_0 Q_E \qquad \text{(Eq. 12.4-7)}$$

where

Q_E = effects of horizontal seismic forces from V, F_{px}, or F_p as specified in Sections 12.8.1, ~~and~~ 12.10, or 13.3.1~~, respectively~~. Where required by ~~in~~ Section 12.5.3 or ~~and~~ 12.5.4, such effects shall result from application of horizontal forces simultaneously in two directions at right angles to each other.

(no change in remainder of Section 12.4.3.1)

12.10.1.1 Diaphragm Design Forces

Floor and roof diaphragms shall be designed to resist design seismic forces from the structural analysis, but shall not be less than that determined in accordance with Eq. 12.10-1 as follows:

$$F_{px} = \frac{\sum_{i=x}^{n} F_i}{\sum_{i=x}^{n} w_i} w_{px} \qquad \text{(Eq. 12.10-1)}$$

where

F_{px} = the diaphragm design force

F_i = the design force applied to Level i

w_i = the weight tributary to Level i,

w_{px} = the weight tributary to the diaphragm at Level x

The force determined from Eq. 12.10-1 shall not be less than:

$$F_{px} = 0.2 S_{DS} I_e w_{px} \qquad \text{(Eq. 12.10-2)}$$

The force determined from Eq. 12.10-1 need not exceed:

$$F_{px} = 0.4S_{DS}I_e w_{px} \quad \text{(Eq. 12.10-3)}$$

$0.4S_{DS}Iw_{px}$ but shall not be less than $0.2S_{DS}Iw_{px}$.

(no change in remainder of Section 12.10.1.1)

12.10.2.1 Collector Elements Requiring Load Combinations with Overstrength Factor for Seismic Design Categories C through F

In structures assigned to Seismic Design Category C, D, E or F, collector elements (see Figure 12.10-1) and their connections, splices, including and their connections to vertical resisting elements shall be designed to resist the maximum of the following:

1. Forces calculated using the seismic load effects including load combinations with overstrength factor of Section 12.4.3 2. with seismic forces determined by the Equivalent Lateral Force procedure of Section12.8 or the Modal Response Spectrum Analysis procedure of Section 12.9.
2. Forces calculated using the seismic load effects including overstrength factor of Section 12.4.3 with seismic forces determined by Equation 12.10-1.
3. Forces calculated using the load combinations of Section 12.4.2.3 with seismic forces determined by Equation 12.10-2.

Transfer forces as described in Section 12.10.1.1 shall be considered.

EXCEPTIONS:

1. The forces calculated above need not exceed those calculated using the load combinations of Section 12.4.2.3 with seismic forces determined by Equation 12.10-3.
2. In structures or portions thereof braced entirely by light-frame shear walls, collector elements and their connections, splices, including and connections to vertical resisting elements need only be designed to resist forces using the load combinations of Section 12.4.2.3 with seismic forces determined in accordance with Section 12.10.1.1.

Analysis and Commentary

In Section 12.4.3.1, it is clarified that diaphragm design forces are earthquake load effects Q_E as used in the load combinations of Section 12.4. Equation numbers in Section 12.10.1.1 are added for the minimum and maximum forces facilitating reference.

Section 12.10.2.1 is revised so that there are three checks that need to be made in determining design forces for collector elements and their connections. ASCE 7-05 did not require that the diaphragm forces of Section 12.10.1-1 be considered.

For structures assigned to SDC C through F, design forces for collector elements and their connections are the maximum of the following:

1. Forces determined from the overall building analysis under the design based shear, V, amplified by the overtstrength factor of Section 12.4.3, that is $\Omega_0 Q_E$;
2. Forces determined from Eq. 12.10-1, F_{px}, amplified by the overstrength factor of Section 12.4.3;
3. Forces determined from Eq. 12.10-2, $F_{px,\,min}$, without any overstrength factor.

The maximum collector forces determined from the above need not exceed those obtained from Eq. 12.10-3, $F_{px\ (max)}$, without overstrength factor.

The reason that the minimum and maximum values of F_{px} are not used with the overstrength factor is that they are independent of the system used, its ductility, and its overstrength. Thus the system overstrength factor is not applied to these values of collector forces, only to collector forces given by Eq. 12.10-1, which uses forces derived from the design base shear (and thus incorporates the Response Modification Coefficient, R).

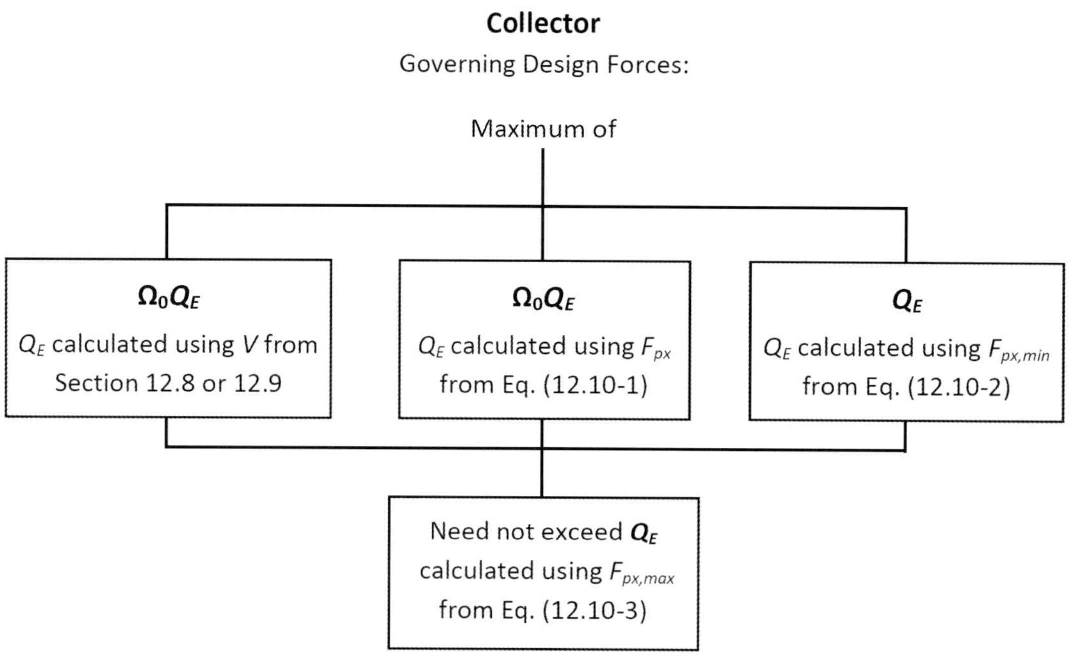

Note: Transfer forces as described in Section 12.10.1.1, if present, need additionally to be considered.

12.11.1 Clarification

Design for Out-of-Plane Forces

At a Glance

It is clarified that the out-of-plane wall design force of Section 12.11.1 is an F_p force.

2010 Standard

12.11.1 Design for Out-of-Plane Forces

Structural walls and their anchorage shall be designed for a force normal to the surface equal to $\underline{F_p =}$ $0.4S_{DS}\underline{I_e}$ times the weight of the structural wall with a minimum force of 10 percent of the weight of the structural wall. Interconnection of structural wall elements and connections to supporting framing systems shall have sufficient ductility, rotational capacity, or sufficient strength to resist shrinkage, thermal changes, and differential foundation settlement when combined with seismic forces.

Analysis and Commentary

In ASCE 7-05, there was no logical path for structural wall forces (out-of-plane) to be included in the seismic load combinations because they were not specifically defined as either V or F_p. The term Q_E as identified under Section 12.4.2.1 is derived only from V or F_p, but the equation for structural wall forces under Section 12.11.1 ($0.4S_{DS}I$) was not labeled as "F_p"; thus it remained technically unresolved how out-of-plane forces entered the load combination equations. This is resolved by stating $F_p = 0.4S_{DS}I_e$ times the weight of the structural wall with a minimum force of 10 percent of the weight of the structural wall.

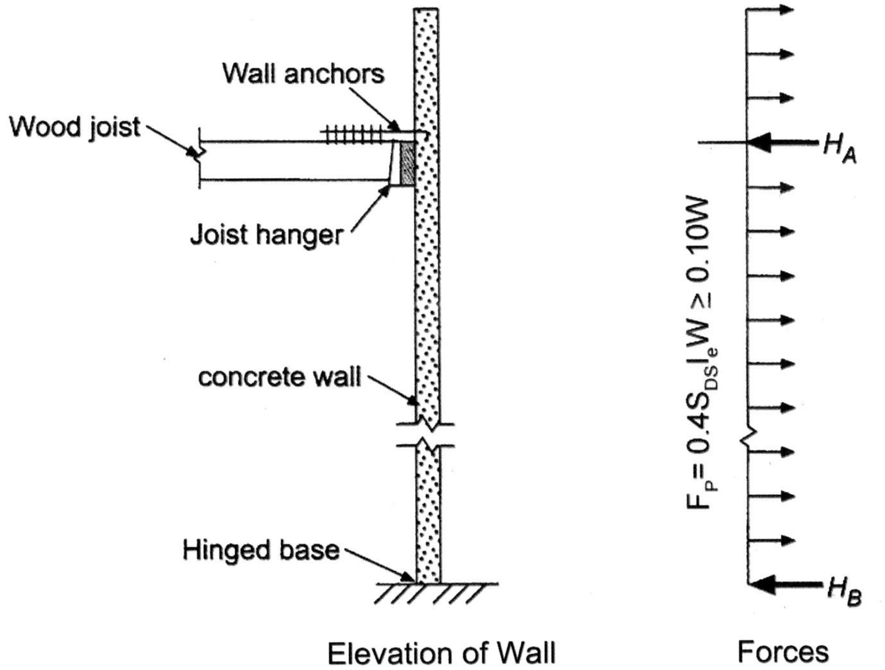

Section 12.11.2 $F_p = 0.4S_{DS}I_eW \geq 0.10W$

11.3, 12.12.3, 12.14.8.5 Addition

Structural Separation and Property Line Setback

At a Glance

Structural separation and setback requirements are added.

2010 Standard

11.3 Symbols

δ_M = maximum inelastic response displacement, considering torsion, Section 12.12.3

δ_{MT} = total separation distance between adjacent structures on the same property, considering torsion, Section 12.12.3

12.12.3 ~~Building~~ Structural Separation

All portions of the structure shall be designed and constructed to act as an integral unit in resisting seismic forces unless separated structurally by a distance sufficient to avoid damaging contact, ~~under total deflection (δ_x) as determined in Section 12.8.6.~~ as set forth in this section.

Separations shall allow for the maximum inelastic response displacement (δ_M). δ_M shall be determined at critical locations with consideration for translational and torsional displacements of the structure including torsional amplifications, where applicable, using the following equation:

$$\delta_M = \frac{C_d \delta_{max}}{I_e} \tag{12.12-1}$$

Where δ_{max} = maximum elastic displacement at the critical location.

Adjacent structures on the same property shall be separated by at least δ_{MT}, determined as follows:

$$\delta_{MT} = \sqrt{(\delta_{M1})^2 + (\delta_{M2})^2} \tag{12.12-2}$$

where δ_{M1} and δ_{M2} are the maximum inelastic response displacements of the adjacent structures at their adjacent edges.

Where a structure adjoins a property line not common to a public way, the structure shall be set back from the property line by at least the displacement δ_M of that structure.

Exception: Smaller separations or property line setbacks are permitted where justified by rational analysis based on inelastic response to design ground motions.

12.14.8.5 Drift Limits and Building Separation

Structural drift need not be calculated. Where a drift value is needed for use in material standards, to determine structural separations between buildings or from property lines, for design of cladding, or for other design requirements, it shall be taken as 1 percent of structural height, h_n, unless computed to be less. All portions of the structure shall be designed to act as an integral unit in resisting seismic forces unless separated structurally by a distance sufficient to avoid damaging contact under the total deflection.

Analysis and Commentary

Structural separation and setback requirements are added to ASCE 7-10. ASCE 7-05 Section 12.12.3 applied only to portions of the same structure. This change expands the scope so that all adjacent structures and setback distances are addressed. A corresponding change is made for the simplified design procedure.

Structural separation and setback provisions were included in the 1997 UBC as well as the 2000 and the 2003 editions of the IBC. However, when the 2006 IBC was being developed, it was decided to delete much of the seismic design provisions (along with much of snow and wind load provisions) from the code itself and incorporate them only through reference to ASCE 7-05. The objective was to eliminate possible error and confusion arising from the IBC's practice of adopting and transcribing, sometimes with modifications, the ASCE 7 provisions. The building separation provisions were also deleted, overlooking the fact that ASCE 7 did not include any such requirements. This error was rectified by having the building separation provisions included in the 2009 IBC by way of a modification to ASCE 7-05. The modification is now incorporated in ASCE 7-10.

The provisions are the same as those included in the 2003 and the 2000 IBC, where the separation between two adjacent buildings needs to be adequate to accommodate the maximum inelastic floor displacements of the two buildings. The maximum inelastic floor displacement is computed as

$$\delta_M = \frac{C_d \delta_{max}}{I_e}$$

where δ_{max} is the maximum elastic displacement that occurs anywhere in a floor from the application of the design base shear to the structure. The maximum elastic displacement, δ_{max}, includes the effects of translation as well as rotation due to torsion (inherent and accidental). It must be noted that δ_{max} is different from the elastic floor displacement, δ_{xe}, which is determined at the center of mass of the floor and is used in ASCE 7-10 Equation 12.8-15 to compute inelastic floor deflection.

The maximum inelastic floor displacements from adjacent buildings are combined by the SRSS (Square Root of Sum of Squares) method to determine the "distance sufficient to avoid damaging contact."

Where a structure adjoins a property line not common to a public way, the structure also needs to be set back from the property line by at least δ_M.

It is also worth noting that the maximum inelastic floor displacement is computed independent of the occupancy of the building as the quantity includes the Importance Factor I_e, which depends on the occupancy, in the denominator. The importance factor gets included in the maximum elastic displacement δ_{max} through the computation of base shear. Thus, when $C_d\delta_{max}$ is divided by I_e, the effect of building occupancy is canceled out.

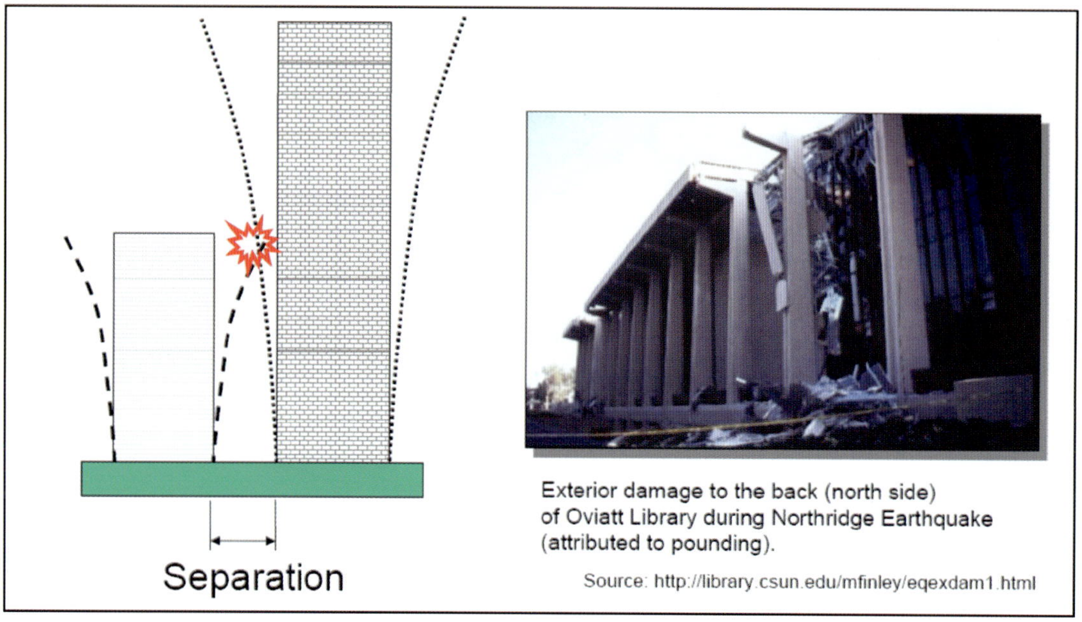

Exterior damage to the back (north side) of Oviatt Library during Northridge Earthquake (attributed to pounding).

Source: http://library.csun.edu/mfinley/eqexdam1.html

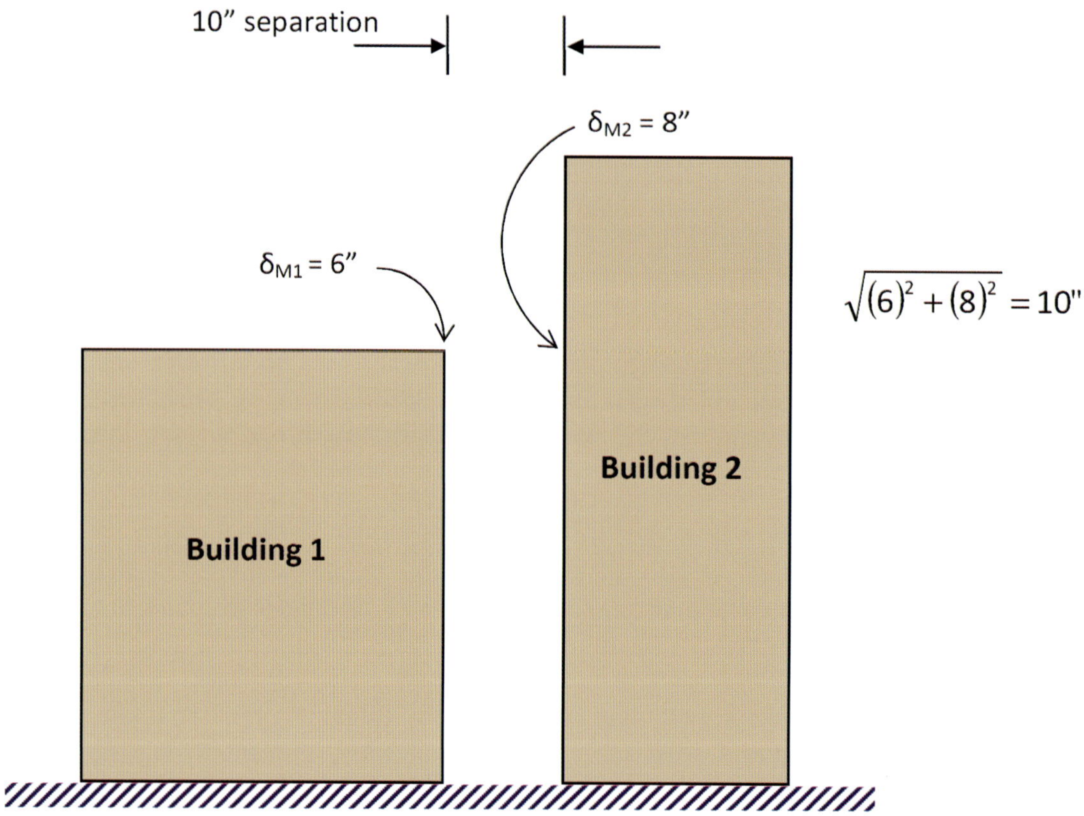

Required Separation for Two Adjacent Buildings

1.4.5, 11.3, 12.11.2, 12.11.2.1, 12.14.7.5 — Modification

Anchorage of Structural Walls and Transfer into Diaphragms

At a Glance

Provisions for the design of anchorage of concrete and masonry walls to diaphragms are modified to be clearer, simpler, and more realistic.

2010 Standard

1.4.5 ~~11.7.5~~ Anchorage of Structural ~~Concrete or Masonry~~ Walls

~~Concrete and masonry~~ ~~W~~walls that provide vertical load bearing or lateral shear resistance for a portion of the structure shall be anchored to the roof and all floors and members that provide lateral support for the wall or that are supported by the wall. The anchorage shall provide a direct connection between the walls and the roof or floor construction. The connections shall be capable of resisting a strength level horizontal force perpendicular to the plane of the wall equal to of 0.2 times the weight of the wall tributary to the connection, but not less than 5 psf (0.24 kN/m[2]) ~~the horizontal forces specified in Section 11.7.3, but not less than a minimum of 280 lb/ linear ft (4.09 kN/m) of wall substituted for E in the load combinations of Section 2.3 or 2.4,~~.

11.3 SYMBOLS

k_a = coefficient defined in Sections 12.11.2 and 12.14.7.5

12.11.2 Anchorage of ~~Concrete or Masonry~~ Structural Walls and Transfer of Design Forces into Diaphragms

12.11.2.1 Wall Anchorage Forces

The anchorage of ~~concrete or masonry~~ structural walls to supporting construction shall provide a direct connection capable of resisting the ~~greater of the~~ following:

$$F_p = 0.4 S_{DS} k_a I_e W_p \qquad (12.11\text{-}1)$$

F_p shall not be taken less than $0.2 k_a I_e W_p$

$$k_a = 1.0 + \frac{L_f}{100} \qquad (12.11\text{-}2)$$

k_a need not be taken larger than 2.0.

~~a. The force set forth in Section 12.11.1~~

~~b. A force of 400 $S_{DS}I$ lb/linear ft (5.84$S_{DS}I$ kN/m) of wall~~

~~c. 280 lb/linear ft (4.09 kN/m) of wall~~

~~Structural walls shall be designed to resist bending between anchors where the anchor spacing exceeds 4 ft (1,219 mm).~~

~~**12.11.2.1 Anchorage of Concrete or Masonry Structural Walls to Flexible Diaphragms.** In addition to the requirements set forth in Section 12.11.2, anchorage of concrete or masonry structural walls to flexible diaphragms in structures assigned to Seismic Design Category C, D, E, or F shall have the strength to develop the out-of-plane force given by Eq. 12.11-1:~~

$$\sout{F_p = 0.8\, S_{DS}\, I\, W_p} \qquad \text{~~(12.11-1)~~}$$

where

F_p = the design force in the individual anchors

S_{DS} = the design spectral response acceleration parameter at short periods per Section 11.4.4

I_e = the ~~occupancy~~ importance factor ~~per~~ determined in accordance with Section 11.5.1

k_a = amplification factor for diaphragm flexibility

L_f = the span, in feet, of a flexible diaphragm that provides the lateral support for the wall; the span is measured between vertical elements that provide lateral support to the diaphragm in the direction considered; use zero for rigid diaphragms

W_p = the weight of the wall tributary to the anchor

Where the anchorage is not located at the roof and all diaphragms are not flexible, the value from Eq. 12.11-1 is permitted to be multiplied by the factor $(1+2z/h)/3$, where z is the height of the anchor above the base of the structure and h is the height of the roof above the base.

Structural walls shall be designed to resist bending between anchors where the anchor spacing exceeds 4 ft (1,219 mm).

12.14.7.5 Anchorage of ~~**Concrete or Masonry**~~ *Structural Walls*

~~Concrete or masonry~~ Sstructural walls shall be anchored to all floors, roofs, and members that provide out-of-plane lateral support for the wall or that are supported by the wall. The anchorage shall provide a positive direct connection between the wall and floor, roof, or supporting member with the strength to resist ~~horizontal forces specified in this section for structures with flexible diaphragms or of Section 13.3.1 (using a_p equal to 1.0 and R_p equal to 2.5) for structures with diaphragms that are not flexible.~~

~~Anchorage of structural walls to flexible diaphragms shall have the strength to develop~~ the out-of-plane force given by Eq. 12.14-10:

$$F_p = 0.4 k_a S_{DS} W_p \qquad (12.14\text{-}10)$$

F_p shall not be taken less than $0.2 k_a W_p$.

$$k_a = 1.0 + \frac{L_f}{100} \qquad (12.14\text{-}11)$$

k_a need not be taken larger than 2.0 where

~~$F_p = 0.8\, S_{DS} W_p$~~ ~~(12.14-10)~~

F_p =the design force in the individual anchors

k_a = amplification factor for diaphragm flexibility

L_f = the span, in feet, of a flexible diaphragm that provides the lateral support for the wall; the span is measured between vertical elements that provide lateral support to the diaphragm in the direction considered; use zero for rigid diaphragms

S_{DS} = the design spectral response acceleration at short periods per Section 12.14.8.1

W_p = the weight of the wall tributary to the anchor

~~**EXCEPTION:** For Seismic Design Category B, the coefficient 0.8 shall be 0.4, with a minimum force of 10 percent of the tributary weight of the wall or $400S_{DS}$ in pounds per foot, whichever is greater.~~

Analysis and Commentary

Several significant changes have been made in the ASCE 7 provisions concerning the design force for the anchorage between walls and floor or roof diaphragms providing lateral support. ASCE 7-05 Section 11.7.5 contained provisions for concrete and masonry walls assigned to SDC A. That section, with modifications, is now Section 1.4.5. The transfer to Chapter 1 has been made because the requirements are considered to be basic structural integrity requirements. The 280 plf minimum requirement has been replaced by 0.2 times the weight of wall tributary to the connection, but not less than 5 psf. Importantly, the requirements apply to all walls, not just concrete and masonry walls.

ASCE 7-05 Sections 12.11.2 (Anchorage of Concrete or Masonry Structural Walls), applicable to structures assigned to SDC B through F, and 12.11.2.1 (Anchorage of Concrete or Masonry Structural Walls to Flexible Diaphragms), applicable to structures assigned to SDC C through F, have been replaced by the newly titled sections, 12.11.2, Anchorage of Structural Walls and Transfer of Design Forces into Diaphragms, and 12.11.2.1, Wall Anchorage Forces, both of which are applicable to structures assigned to SDC B through F. The new section titles improve the clarity and organization of the anchorage provisions. Similar revisions are made to the anchorage provisions in Section 12.14 for the simplified seismic design method.

There are significant substantive changes to the anchorage provisions. First, there is no longer any distinction between concrete and masonry walls and all walls. Second, the lower-bound anchorage force of $0.10W_p$ (280 plf in the case of concrete and masonry walls) is replaced by a minimum force of $0.2k_a I_e W_p$. As clearly indicated by the following comparison figures, this results in rather significant increases in the anchorage design force for taller walls in areas of moderate to low seismic hazard (where S_{DS} values are moderate to low). Third, the anchorage design force for walls supported by flexible diaphragms used to be twice that for walls supported by rigid diaphragms. ASCE 7-10 provides a gradual increase in anchorage design force through a multiplier k_a, which increases from 1.0 to 2.0, as the flexible diaphragm span increases from zero to 100 ft or more. In a further important change, where the anchorage is not located at the roof and all diaphragms are not flexible, the anchorage design force given by Eq. 12.11-1 may be reduced by $(1 + 2z/h)/3$, where z is the height of the anchor above the base of the structure and h is the height of the roof above the base. This is consistent with the variation in seismic design force for nonstructural components attached to a building, along the height of the building, as given in Section 13.3.1.

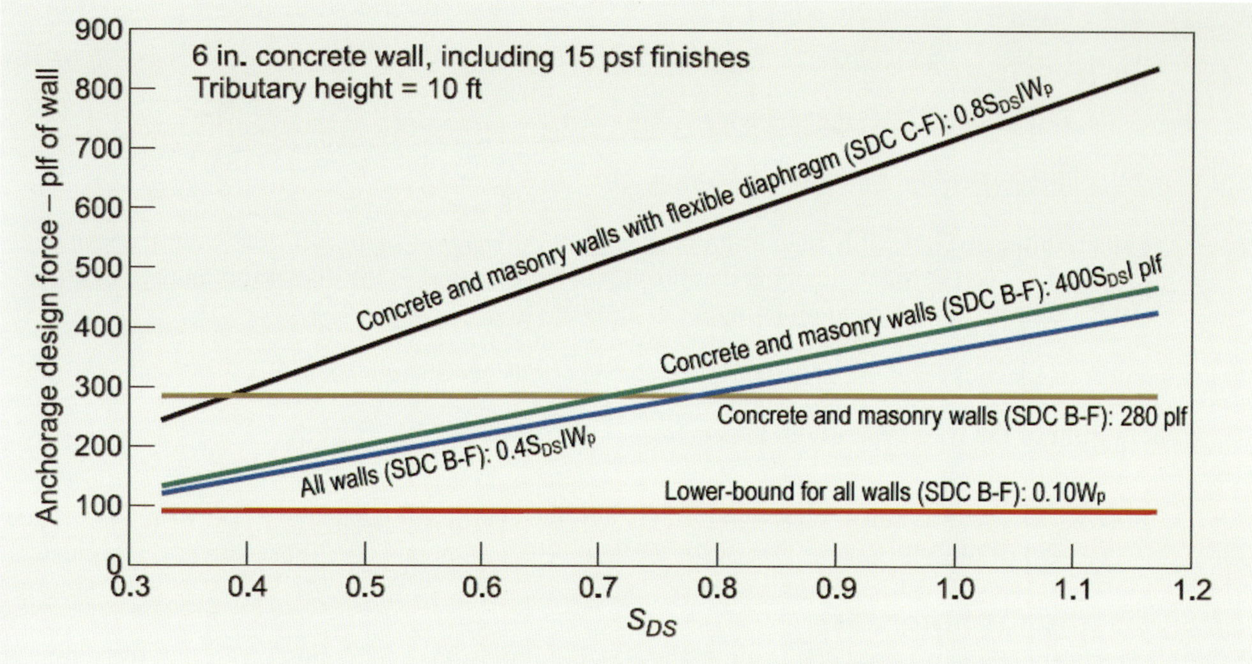

ASCE 7-05 Anchorage design force (tributary wall height = 10 ft)
Courtesy: International Code Council

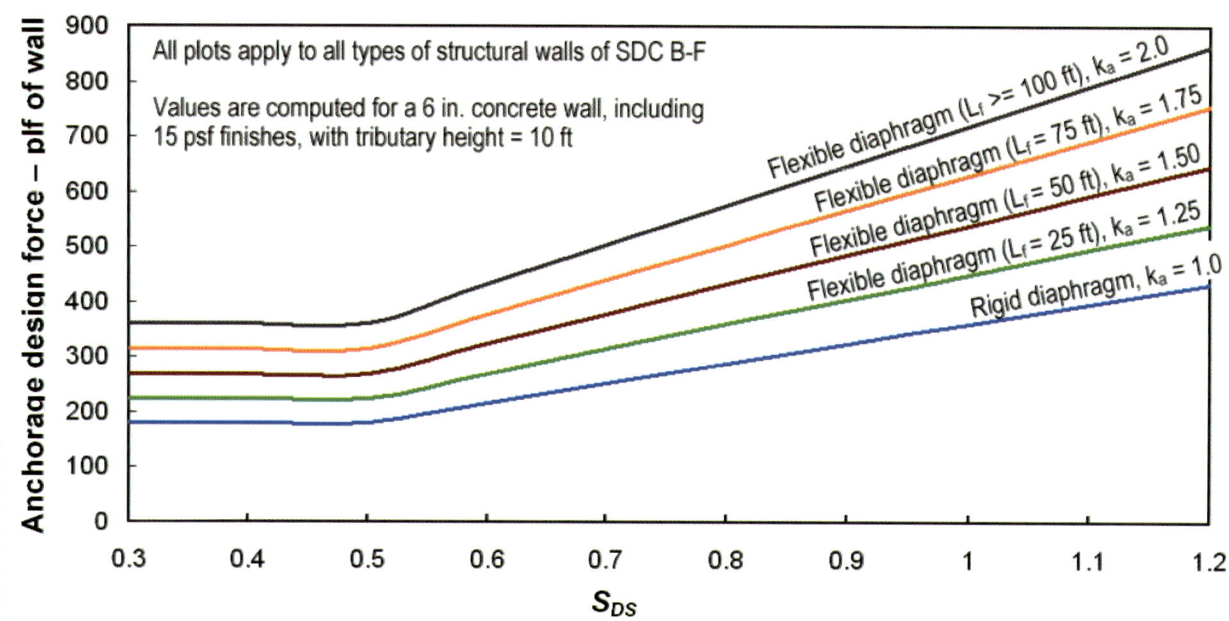

ASCE 7-10 Anchorage design force (tributary wall height = 10 ft)
Courtesy: S.K. Ghosh Associates Inc.

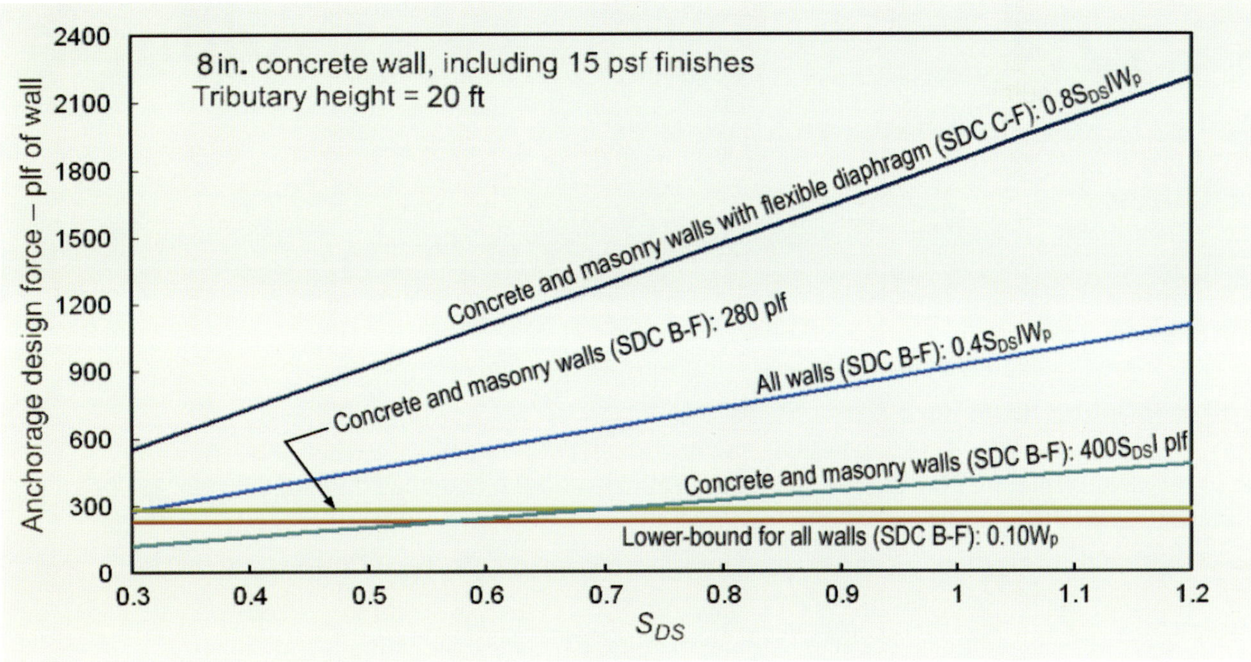

ASCE 7-05 Anchorage design force (tributary wall height = 20 ft)
Courtesy: S.K. Ghosh Associates Inc.

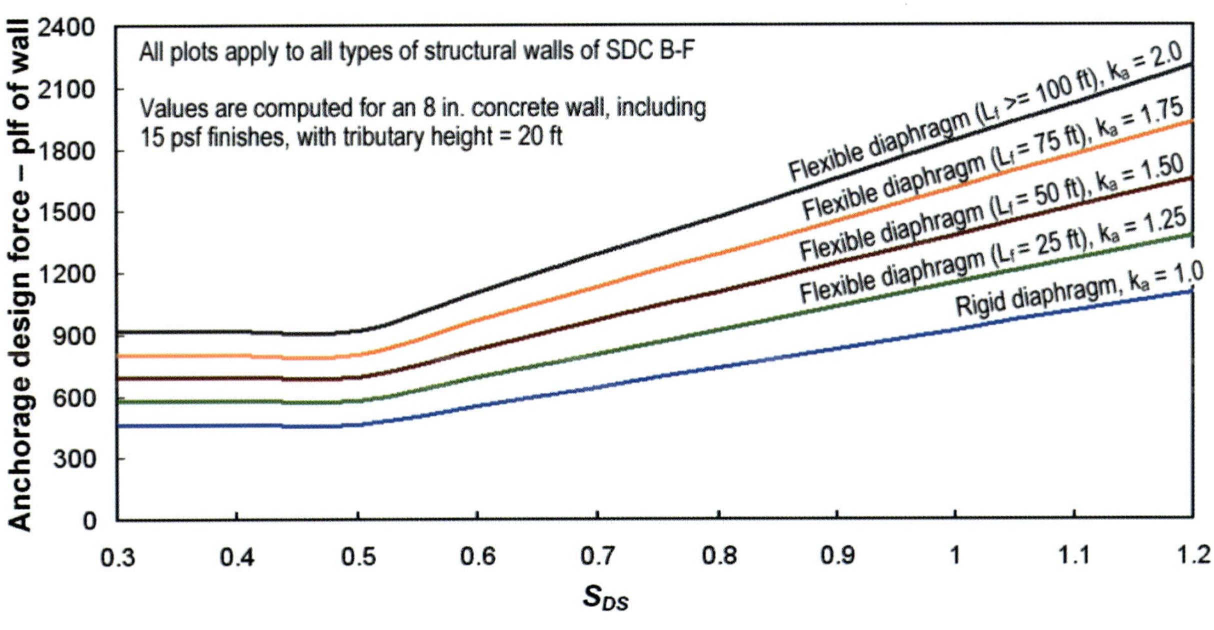

ASCE 7-10 Anchorage design force (tributary wall height = 20 ft)
Courtesy: S.K. Ghosh Associates Inc.

12.12.4 Addition

Members Spanning between Structures

At a Glance

New requirements are added to address design for maximum anticipated relative displacements of members spanning between structures.

2010 Standard

12.12.4 Members Spanning between Structures

Gravity connections or supports for members spanning between structures or seismically separate portions of structures shall be designed for the maximum anticipated relative displacements. These displacements shall be calculated:

1. Using the deflection calculated at the locations of support, per Eq. 12.8-15 multiplied by 1.5 R/C_d, and
2. Considering additional deflection due to diaphragm rotation including the torsional amplification factor calculated per Section 12.8.4.3 where either structure is torsionally irregular, and
3. Considering diaphragm deformations, and
4. Assuming the two structures are moving in opposite directions and using the absolute sum of the displacements.

Analysis and Commentary

ASCE 7-05 provisions did not specifically address the common condition where a seismic separation exists but the gravity system is not separate. In such a case large relative movements of the seismically separate building portions may lead to loss of gravity support for members that bridge between the two portions unless supports are designed to accommodate such displacements. Five requirements are given for estimating these movements.

1. *Using the deflection calculated at the center of mass per Eq. 12.8-15 multiplied by* $1.5R/C_d$*;*
   The base shear used in calculating deflections in Eq. 12.8-15 corresponds to 2/3 of MCE; the 1.5 factor adjusts it to an MCE-level force. Use of values of C_d less than R underestimates deflections according to Uang and Maarouf, 1994 (Journal of Structural Engineering, ASCE, Vol. 120, No. 8, pp. 2423-2436). The R/C_d adjustment accounts for that.
2. *Considering additional deflection due to diaphragm rotation;*
   Eq. 12.8-15 uses center-of-mass displacements. Diaphragm rotation can add significantly to these.
3. *Using the torsional amplification factor calculated per Section 12.8.4.3 when either structure is torsionally irregular;*
   Additional rotation for torsionally irregular structures should be included because of the high consequence of loss of gravity support.
4. *Considering diaphragm deformations;*
   In some types of construction the displacement of the diaphragm can be considerable.

5. *Assuming the two portions of the building are moving in opposite directions and using the absolute sum of displacements;*
   The absolute sum is used instead of a modal combination, which would represent a probable maximum, because of the high consequence of loss of gravity support.

Example of Bridge Spanning Between Buildings

Courtesy: Daniel Schwen, commons.wikimedia.org, under the terms of GNU Free Documentation License, Version 1.2 or later

12.14.7.2, Chapter 23

Modification and Addition

Openings or Reentrant Building Corners

At a Glance

Type II shear walls complying with AISI S213 are now exempt from the requirement that openings in shear walls be provided with reinforcement at the edges of the openings.

2010 Standard

12.14.7.2 Openings or Reentrant Building Corners

Except where as otherwise specifically provided for in this standard, openings in shear walls, diaphragms, or other plate-type elements, shall be provided with reinforcement at the edges of the openings or reentrant corners designed to transfer the stresses into the structure. The edge reinforcement shall extend into the body of the wall or diaphragm a distance sufficient to develop the force in the reinforcement.

EXCEPTION: ~~Perforated s~~Shear walls of wood structural panels are permitted where designed in accordance with AF&PA SDPWS for perforated shear walls or AISI S213 for Type II shear walls.

Chapter 23
SEISMIC DESIGN REFERENCE DOCUMENTS

ANSI/AISI S213 with S1-09 ~~Lateral~~
Sections 12.14.7.2, 14.1.1, 14.1.2, 14.1.4.2
North American Standard for the Design of Cold Formed Steel Framing, Lateral Design, 2007~~04~~, with Supplement 1, 2009

Analysis and Commentary

Perforated shear walls are permitted in the AF&PA *Special Design Provisions for Wind and Seismic* and are referenced in the exception to ASCE 7-05 Section 12.14.7.2. A similar system called "Type II shear walls" is permitted for cold formed steel framing in the AISI S213 standard. The exception is expanded to recognize Type II shear walls based on testing that has shown that a portion of the shear forces can be transferred through the steel framing around openings.

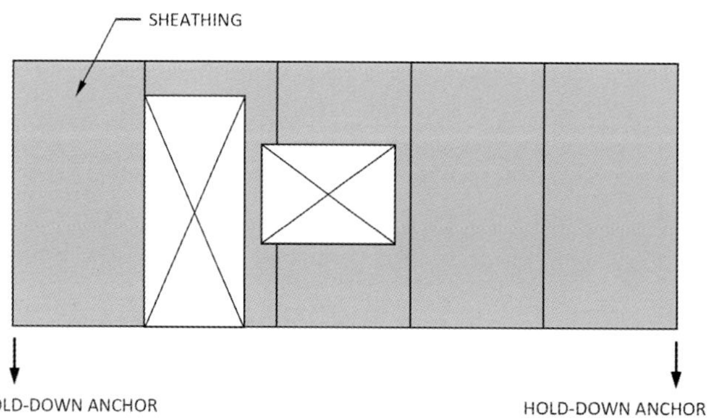

Illustration of Type II Shear Wall per AISI S213

Part IV
Chapter 13 Seismic Design Requirements for Nonstructural Components

13.1.3 Modification

Component Importance Factor

At a Glance

Egress stairways are identified as a component requiring an $I_p = 1.5$ assignment.

2010 Standard

13.1.3 Component Importance Factor

All components shall be assigned a component importance factor as indicated in this section. The component importance factor, I_p, shall be taken as 1.5 if any of the following conditions apply:

1. The component is required to function for life-safety purposes after an earthquake, including fire protection sprinkler systems and egress stairways.
2. The component conveys, supports, or otherwise contains toxic, highly toxic, or explosive substances where the quantity of the material exceeds a threshold quantity established by the authority having jurisdiction and is sufficient to pose a threat to the public if released.
3. The component is in or attached to an Occupancy Category IV structure and it is needed for continued operation of the facility or its failure could impair the continued operation of the facility.
4. The component conveys, supports, or otherwise contains hazardous substances and is attached to a structure or portion thereof classified by the authority having jurisdiction as a hazardous occupancy.

All other components shall be assigned a component importance factor, I_p, equal to 1.0.

Table 13.5-1 Coefficients For Architectural Components

Architectural Component or Element	a_p [a]	R_p [b]
Egress stairways not part of the building structure	1.0	2.5
(remainder of table not shown)		

Analysis and Commentary

Requiring the application of a component importance factor of 1.5 for components required to function for life-safety purposes after an earthquake relies on the judgment of designers and authorities having jurisdiction to determine which components serve such a function. There are certain components, however, that clearly serve a life-safety function wherever they are installed. Automatic fire protection sprinkler systems are one example and they are currently listed in Item 1 of Section 13.1.3. Exit stairways are another example. Occupants rely on egress stairways to safely exit from buildings after earthquakes.

Example of Egress Stairway Not Part of Building Structure

13.1.4, Table 13.5-1

Modification and Clarification

Exemptions for Nonstructural Components and Coefficients for Architectural Components

At a Glance

Furniture and temporary and moveable equipment are added to list of exceptions and improved language is provided to clarify intent of exceptions. One important change is made to Table 13.5-1, Coefficients for Architectural Components.

2010 Standard

13.1.4 Exemptions

The following nonstructural components are exempt from the requirements of this section:

1. Furniture (except storage cabinets as noted in Table 13.5-1).
2. Temporary or moveable equipment.
3. ~~1.~~Architectural components in Seismic Design Category B other than parapets supported by bearing walls or shear walls provided that the component importance factor, I_p, is equal to 1.0.
4. ~~2.~~ Mechanical and electrical components in Seismic Design Category B.
5. ~~3.~~ Mechanical and electrical components in Seismic Design Category C provided that the component importance factor, I_p, is equal to 1.0.
6. ~~4.~~ Mechanical and electrical components in Seismic Design Category D, E, or F where all of the following apply:
   a. tThe component importance factor, I_p, is equal to 1.0 ~~and both the following conditions apply:~~
   b. The component is positively attached to the structure;
   c.~~a.~~ Flexible connections are provided between the components and associated ductwork, piping, and conduit ~~are provided~~; and either
      i. The component weighs 400 lb (1,780 N) or less and has a center of mass located 4 ft (1.22m) or less above the adjacent floor level; or ~~b. Components are mounted 4 ft (1.22 m) or less above a floor level and weigh 400 lb (1780 N) or less.~~
      ii. The component weighs 20 lb (89 N) or less or, in the case of a distributed system, 5 lb/ft (73 N/m) or less.

~~5. Mechanical and electrical components in Seismic Design Category D, E, and F where the component importance factor I_p, is equal to 1.0 and both the following conditions apply:~~
~~a. Flexible connections between the components and associated ductwork, piping, and conduit are provided, and~~
~~b. The components weigh 20 lb (89 N) or less, for distribution systems, weighing 5 lb/ft (73 N/m) or less.~~

Table 13.5-1 Coefficients for Architectural Components

Architectural Component	$a_p{}^a$	$R_p{}^b$
Cabinets		
Permanent floor-supported ~~S~~storage cabinets ~~and laboratory equipment (including contents)~~ over 6 ft (1,829 mm) tall, including contents	1.0	2.5
Permanent floor-supported library shelving, book stacks and bookshelves ~~(including contents)~~ over 6 ft (1,829 mm) tall, including contents	1.0	2.5
Laboratory equipment	1.0	2.5

Analysis and Commentary

Two new exceptions are added to Section 13.1.4. This section is intended to exempt certain nonstructural components from requiring seismic anchorage. The two new exceptions are meant to cover items that have typically been understood or treated as exempt but that have never been explicitly exempted in building codes.

Improved language has been incorporated into newly numbered Exception No. 6 to clarify the necessary conditions for mechanical and electrical equipment to be exempt from the requirements of this section. The ASCE 7-05 language was not clear with respect to the height/mass exception. Exception No. 6 now is as follows:

1. The $I_p = 1.0$ requirement remains.
2. All components are expected to be positively attached to the structure (no free-sliding equipment).
3. Flexible connections are required in all cases.
4. No exemptions for equipment weighing more than 400 lb. with a center of mass more than 4 feet above the adjacent floor level.
5. No exemptions for equipment with a center of mass 4 feet or more above the adjacent floor level unless it weighs less than 20 lb or 5 lb/ft.

The changes to Table 13.5-1 are intended to clarify that storage cabinets require anchorage only when they are over 6 feet tall.

Permanent floor-supported storage cabinets over 6 feet tall are subject to the requirements of Chapter 13.
Courtesy: Federal Emergency Management Agency

11.2, 13.1.6, 13.1.7, 13.4, Table 13.5-1, 13.6.1, 13.6.2

Modification and Addition

Definitions and Usage of Terms Related to Chapter 13, Including Friction Clip

At a Glance

Consistency and clarity are provided in the usage of the terms *component, nonstructural component, support, and attachment.* The definition for DESIGNATED SEISMIC SYSTEM is revised to correspond with the 2006 and 2009 IBC. A new definition for FRICTION CLIP is introduced and the friction clip provision is revised. Determination of component period is clarified.

2010 Standard

11.2 DEFINITIONS

The following definitions apply only to the seismic requirements of this standard.

ATTACHMENTS: Means by which nonstructural components or supports of nonstructural components ~~and their supports~~ are secured or connected to the seismic force-resisting system of the structure. Such attachments include anchor bolts, welded connections, and mechanical fasteners.

COMPONENT: A part ~~or element~~ of an architectural, electrical, or mechanical ~~or structural~~ system.

Component, ~~Equipment~~ Nonstructural: A part of an architectural, mechanical or electrical ~~component or element that is part of a mechanical and/or electrical~~ system within or without a building or nonbuilding structure ~~system~~.

Component, Flexible: ~~C~~Nonstructural component~~, including its attachments,~~ having a fundamental period greater than 0.06 s.

Component, Rigid: ~~C~~Nonstructural component~~, including its attachments~~ having a fundamental period less than or equal to 0.06 s.

~~**COMPONENT SUPPORT:** Those structural members or assemblies of members, including braces, frames, struts, and attachments that transmit all loads and forces between systems, components or elements and the structure.~~

DESIGNATED SEISMIC SYSTEMS: ~~The seismic force-resisting system and t~~Those nonstructural ~~architectural, electrical, and mechanical systems or their~~ components that require design in accordance with Chapter 13 and for which the component importance factor, I_p, is greater than 1.0.

FLEXIBLE ~~EQUIPMENT~~ CONNECTIONS: Those connections between equipment components that permit rotational and/or translational movement without degradation of performance. Examples include universal joints, bellows expansion joints, and flexible metal hose.

FRICTION CLIP: A device that relies on friction to resist applied loads in one or more directions to anchor a nonstructural component. Friction is provided mechanically and is not due to gravity loads.

SUPPORTS: Those ~~structural~~ members, assemblies of members, or manufactured elements, including braces, frames, legs, lugs, snubbers, hangers, saddles, or struts, and associated fasteners that ~~which~~ transmit loads between ~~the~~ nonstructural components and their attachments to the structure.

13.1.6 Reference Documents

Where a reference document provides a basis for the earthquake-resistant design of a particular type of nonstructural ~~system or~~ component, that document is permitted to be used, subject to the approval of the authority having jurisdiction and the following conditions:

a. The design earthquake forces shall not be less than those determined in accordance with Section 13.3.1.
b. Each nonstructural component's seismic interactions with all other connected components and with the supporting structure shall be accounted for in the design. The component shall accommodate drifts, deflections, and relative displacements determined in accordance with the applicable seismic requirements of this standard.
c. Nonstructural component anchorage requirements shall not be less than those specified in Section 13.4.

13.1.7 Reference Documents Using Allowable Stress Design

Where a reference document provides a basis for the earthquake-resistant design of a particular type of ~~system or~~ component, and the same reference document defines acceptance criteria in terms of allowable stresses rather than strengths, that reference document is permitted to be used. The allowable stress load combination shall consider dead, live, operating, and earthquake loads in addition to those in the reference document. The earthquake loads determined in accordance with Section 13.3.1 shall be multiplied by a factor of 0.7. The allowable stress design load combinations of Section 2.4 need not be used. The component ~~or system~~ shall also accommodate the relative displacements specified in Section 13.3.2.

13.4 NONSTRUCTURAL COMPONENT ANCHORAGE

Nonstructural c~~C~~omponents and their supports shall be attached (or anchored) to the structure in accordance with the requirements of this section and the attachment shall satisfy the requirements for the parent material as set forth elsewhere in this standard.
(no change in the remainder of Section 13.4)

13.4.6 Friction Clips

Friction clips in Seismic Design Categories D, E, or F shall not be used for ~~anchorage attachment.~~ supporting sustained loads in addition to resisting seismic forces. C-type beam and large flange clamps are permitted for hangers provided they are equipped with restraining straps equivalent to those specified in NFPA 13, Section 9.3.7. Lock nuts or equivalent shall be provided to prevent loosening of threaded connections.

Table 13.5-1 Coefficients For Architectural Components

Architectural Component ~~or Element~~	$a_p{}^a$	$R_p{}^b$

(No changes to table entries)

[a]A lower value for a_p shall not be used unless justified by detailed dynamic analysis. The value for a_p shall not be less than 1.00. The value of $a_p = 1$ is for rigid components and rigidly attached components. The value of $a_p = 2.5$ is for flexible components and flexibly attached components. ~~See Section 11.2 for definitions of rigid and flexible.~~

13.6.1 General

Mechanical and electrical components and their supports shall satisfy the requirements of this section. The attachment of mechanical and electrical components and their supports to the structure shall meet the requirements of Section 13.4. Appropriate coefficients shall be selected from Table 13.6-1.

EXCEPTION: *(no change to exception)*

Where design of mechanical and electrical components for seismic effects is required, consideration shall be given to the dynamic effects of the components, their contents, and where appropriate, their supports and attachments. In such cases, the interaction between the components and the supporting structures, including other mechanical and electrical components, shall also be considered.

13.6.2 Component Period

The fundamental period of the nonstructural ~~mechanical and electrical~~ component (including ~~and~~ its supports and attachment to the structure ~~building~~), T_p, shall be determined by the following equation provided that the component, supports, and attachment can be reasonably represented analytically by a simple spring and mass single degree-of-freedom system:

$$T_p = 2\pi\sqrt{\frac{W_p}{K_p g}} \qquad (13\text{-}6.1)$$

where

T_p = component fundamental period

W_p = component operating weight

g = gravitational acceleration

K_p = combined stiffness ~~of resilient support system~~ of the component, supports and attachments, determined in terms of load per unit deflection at the center of gravity of the component

Alternatively, the fundamental period of the component, T_p, in s is permitted to be determined from experimental test data or by a properly substantiated analysis.

Analysis and Commentary

Consistency and clarity are provided in the usage of the terms *component, nonstructural component, support, and attachment.* The distinction among attachments, components, and supports is necessary to the understanding of the requirements for nonstructural components and nonbuilding structures in Chapters 13 and 15, respectively. Common cases associated with nonstructural components are illustrated in Commentary Figure C11-1 in ASCE 7-10 and reproduced herein. Note that the definitions of components, supports, and attachments are generally applicable to components with a defined envelope in the as-manufactured condition and for which additional supports and attachments are required to provide support in the as-built condition. As the ASCE 7-10 Commentary to Section 11.2 points out, this distinction may not always be clear, particularly when the component is equipped with prefabricated supports; therefore, judgment must be used in the assignment of forces to specific elements in accordance with the provisions of Chapter 13.

The term designated seismic system is no longer used outside of Chapter 13, Seismic Design Requirements for Nonstructural Components. This revision in the definition of the term retains the meaning of the term as used in Chapter 13 and harmonizes ASCE 7 and the Chapter 13 definition with that found in Section 1702 of the 2006 and 2009 IBC.

The term *friction clip* was used but not defined in ASCE 7-05 and could be interpreted to include many common types of devices used for the anchorage of nonstructural components, such as cable clamps, wedge anchors, cold-formed metal channel (strut) connections, restrainer clips for suspended ceiling tiles, etc. Since there was no definition, it was not clear whether the "friction clips" would be allowed to resist forces in directions other than the direction(s) in which friction was provided. The term *friction clip* is now defined in Section 11.2 in a general way to encompass C-type beam clamps as well as cold-formed metal channel (strut) connections. Friction clips are suitable to resist seismic forces provided they are properly designed and installed, but under no circumstances should they be relied upon to resist sustained gravity loads.

Section 13.4.6 is revised so that the prohibition of friction connections is limited to those that can dislodge due to earthquake shaking and cause the nonstructural component to displace and cause a life safety hazard or, in the case of Designated Seismic Systems, in a loss of operability. Thus, nonstructural components are prevented from relying on friction connections to provide gravity support, since these connections can be dislodged and create a falling hazard. Friction connections, using a mechanical means and not relying on gravity to generate friction, are allowed for resisting lateral loads where gravity support for the component can be maintained with the loss of the friction connection. Friction connections that are required to resist both gravity loads and lateral loads are prohibited, to ensure that failure of the friction connection due to lateral loads does not affect the gravity support for the component.

The modification to Footnote a of Table 13.5-1 reflects the fact that the a_p and R_p values in the table are based on judgment and in aggregate reflect what is judged to be the correct coefficient for the design of the listed components. In general, it is not envisioned that a period calculation would be performed for architectural components to establish whether they are flexible or rigid.

Language is also altered in Section 13.6.2 specifying the need to consider the structural response for the determination of the component period where the component and supporting structure have a significant dynamic interaction, such as might be the case for a large roof-mounted chiller unit.

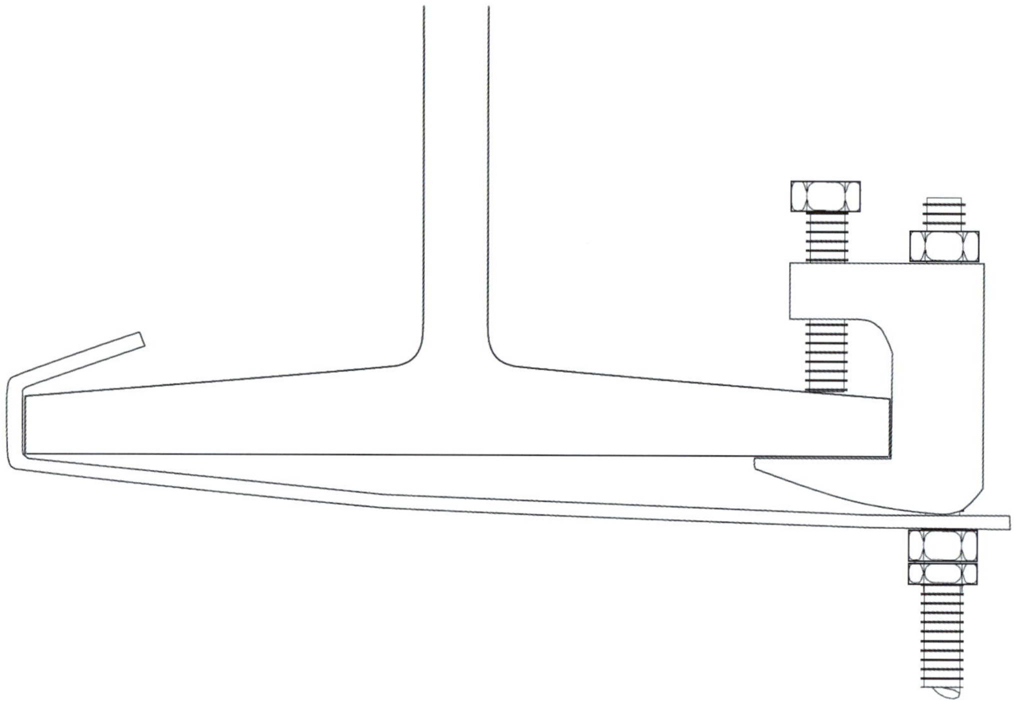

C-type Beam Clamp Equipped with a Restraining Strap

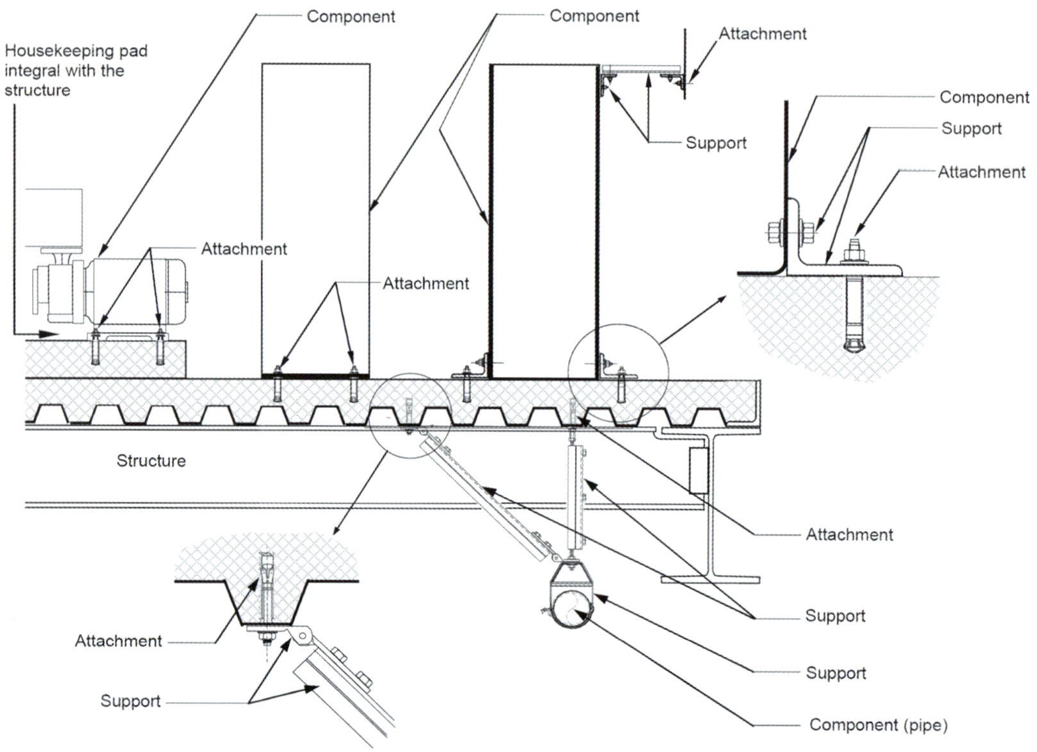

Examples of Components, Supports, and Attachments

13.1.6 Addition

Reference Documents

At a Glance

A new requirement for compliance with Section 13.4 Nonstructural Component Anchorage is added where a reference document is used for seismic design of a nonstructural component.

2010 Standard

13.1.6 Reference Documents

Where a reference document provides a basis for the earthquake-resistant design of a particular type of nonstructural component, that document is permitted to be used, subject to the approval of the authority having jurisdiction and the following conditions:

a. The design earthquake forces shall not be less than those determined in accordance with Section 13.3.1.

b. Each component's seismic interactions with all other connected components and with the supporting structure shall be accounted for in the design. The component shall accommodate drifts, deflections, and relative displacements determined in accordance with the applicable seismic requirements of this standard.

c. Nonstructural component anchorage requirements shall not be less than those specified in Section 13.4.

Analysis and Commentary

Reference documents can be excellent resources for the seismic design of nonstructural components. However, implementation of earthquake-resistant design in reference documents may be behind the ASCE 7 provisions. Design forces are kept current by reference to Section 13.3.1 in Item (a) of Section 13.1.6. The new reference to Section 13.4 provides assurance that connection of the nonstructural component to the structure is adequate. Anchorage that is required by the reference standard needs to be held to the same degree of care as would be attained by using ASCE 7 provisions. Note that Section 13.6.11 Other Mechanical and Electrical Components requires compliance with Section 13.4.

CHECKLIST FOR USING A REFERENCE DOCUMENT for SEISMIC DESIGN of a NONSTRUCTURAL COMPONENT:

- ☑ Approval of Authority Having Jurisdiction
- ☑ Seismic Design Force ≥ F_p in Section 13.3.1
- ☑ Compliance with Section 13.3.2 Seismic Relative Displacements
- ☑ Compliance with Section 13.4 Nonstructural Component Anchorage

13.2.1 and 13.2.2 Modification

Applicable Requirements for Architectural, Mechanical, and Electrical Components, Supports, and Attachments; Special Certification Requirements for Designated Seismic Systems

At a Glance

Clarification and improved consistency in language is provided in applicable requirements for architectural, mechanical, and electrical components, supports, and attachments. Special certification is required to be provided by manufacturer rather than supplier; method of certification for active parts or energized components is further defined.

2010 Standard

13.2.1 Applicable Requirements for Architectural, Mechanical, and Electrical Components, Supports, and Attachments

Architectural, mechanical, and electrical components, supports, and attachments shall comply with the sections referenced in Table 13.2-1. These requirements shall be satisfied by one of the following methods:

1. Project-specific design and documentation prepared and submitted for approval to the authority having jurisdiction after review and acceptance by a registered design professional.
2. Submittal of the manufacturer's certification that the component is seismically qualified by at least one of the following:
   a. Analysis~~.~~, or
   b. Testing in accordance with the alternative set forth in Section 13.2.5~~.~~, or
   c. Experience data in accordance with the alternative set forth in Section 13.2.6.

13.2.2 Special Certification Requirements for Designated Seismic Systems

Certifications shall be provided for designated seismic systems assigned to Seismic Design Categories C through F as follows:

1.~~a.~~ Active mechanical and electrical equipment that must remain operable following the design earthquake ground motion shall be certified by the manufacturer ~~supplier~~ as operable whereby active parts or energized components shall be certified ~~based~~ exclusively on the basis of approved shake table testing in accordance with Section 13.2.5 or experience data in accordance with Section 13.2.6 unless it can be shown that the component is inherently rugged by comparison with similar seismically qualified components. Evidence demonstrating compliance with ~~of~~ this requirement shall be submitted for approval to the authority having jurisdiction after review and ~~approval~~ acceptance by a ~~the~~ registered design professional.

2.~~b.~~ Components with hazardous substances and assigned a component importance factor, I_p, of 1.5 in accordance with Section 13.1.3 shall be certified by the manufacturer ~~supplier~~ as maintaining containment following the design earthquake ground motion by (1) analysis, (2) approved shake table testing in accordance with Section 13.2.5, or (3) experience data in accordance with Section 13.2.6. Evidence demonstrating compliance ~~of~~ with this requirement shall be submitted for approval to the authority having jurisdiction after review and ~~approval~~ acceptance by a ~~the~~ registered design professional.

Analysis and Commentary

In Items 1 and 2 of Section 13.2.2, "approval" is changed to "acceptance," which is a more appropriate term for the responsibilities of a registered design professional. "Approval" is typically the prerogative of the authority having jurisdiction. This change will also make the items consistent with other provisions of ASCE 7 that specify acceptance by a registered design professional.

In Section 13.2.2, the phrase "the registered design professional" is changed to "a registered design professional" in two places. When ASCE 7 intends to be more specific than the way it was in ASCE 7-05, it typically uses the phrase: the registered design professional responsible for the structural design, or something to that effect, as in Sections 17.2.4.9 and 18.2.5.4. Here that is not the intent. So ASCE 7 has chosen to be less specific by going from "the" to "a" registered design professional, meaning any registered design professional.

Language is added stating that active parts or energized components be certified by approved shake table testing or experience data unless it can be shown that the component is inherently rugged by comparison with similar seismically qualified components. Active equipment has parts that rotate, move mechanically or are energized during operation. Evaluating post-earthquake operational performance for active equipment by analysis generally involves sophisticated modeling with experimental validation and may not be reliable. Therefore, the use of analysis for active or energized components is not permitted unless a comparison can be made to components that have otherwise been deemed as rugged.

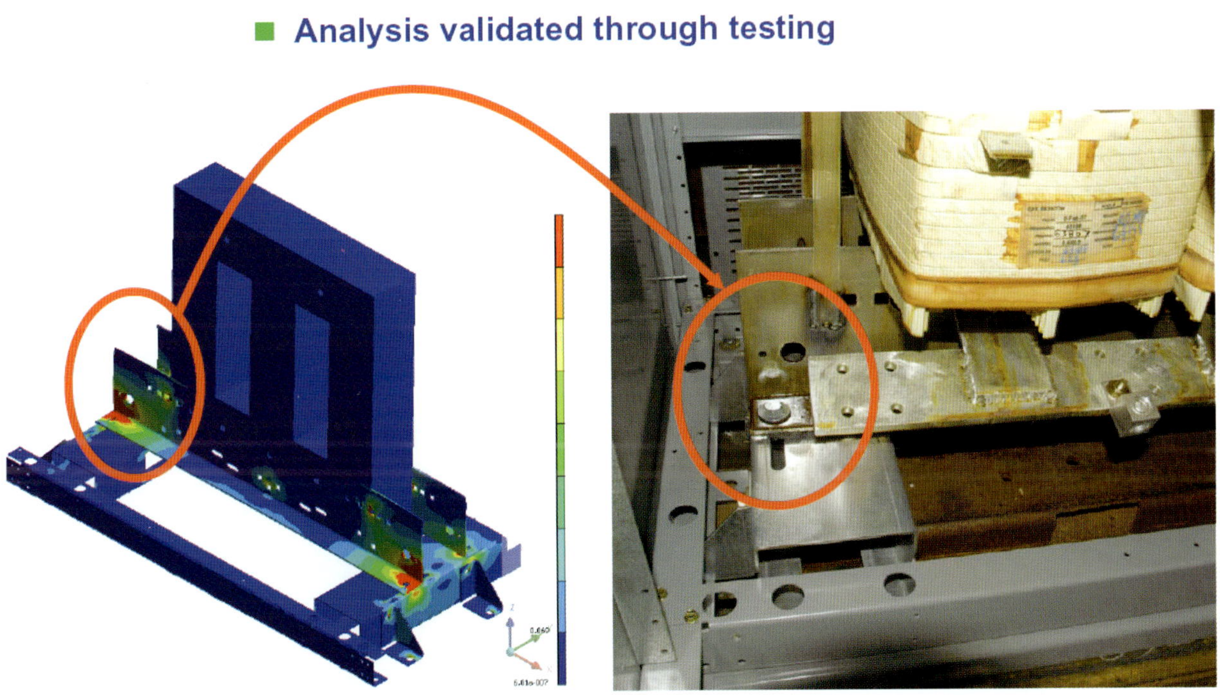

Testing in Accordance with Section 13.2.5 as Method of Certification by Manufacturer
Courtesy: Philip Caldwell

11.3, 13.3.2, 13.3.2.2 Modification

Seismic Relative Displacement

At a Glance

The Importance Factor I_e is included in the equation for the determination of seismic relative displacement between two structures.

2010 Standard

11.3 SYMBOLS

D_{pI} = seismic relative ~~seismic~~ displacement; see ~~that a component must be designed to accommodate as defined in~~ Section 13.3.2

13.3.2 Seismic Relative Displacements

The effects of seismic relative displacements shall be considered in combination with displacements caused by other loads as appropriate. Seismic relative displacements, D_{pI}, ~~(D_p)~~ shall be ~~determined in accordance with the equations set forth in Sections 13.3.2.1 and 13.3.2.2~~ determined in accordance with Eq. 13.3-5 as

$$D_{pI} = D_p I_e \qquad (13.3\text{-}5)$$

where

I_e = the importance factor in Section 11.5.1.

D_p = displacement determined in accordance with the equations set forth in Sections 13.3.2.1 and 13.3.2.2.

Section 13.3.2.2 Displacements between Structures

For two connection points on separate Structures A and B or separate structural systems, one at a height h_x and the other at a height h_y, D_p shall be determined as:

$$D_p = |\delta_{xA}| + |\delta_{yB}| \qquad \text{(Eq. 13.3-}\sout{7}\underline{8})$$

D_p is not required to be taken as greater than:

$$D_p = \frac{h_x \Delta_{aA}}{h_{sx}} + \frac{h_y \Delta_{aB}}{h_{sx}} \qquad \text{(Eq. 13.3-}\sout{8}\underline{9})$$

where

D_p = relative seismic displacement that the component must be designed to accommodate

δ_{xA} = deflection at building Level x of Structure A, determined ~~by an elastic analysis as defined in Section 12.8.6~~ in accordance with Eq. (12.8-15)

δ_{yA} = deflection at building Level y of Structure A, determined ~~by an elastic analysis as defined in Section 12.8.6~~ in accordance with Eq. (12.8-15).

δ_{yB} = deflection at building Level y of Structure B, determined ~~by an elastic analysis as defined in Section 12.8.6~~ in accordance with Eq. (12.8-15).

h_x = height of Level x to which upper connection point is attached

h_y = height of Level y to which lower connection point is attached

Δ_{aA} = allowable story drift for Structure A as defined in Table 12.12-1

Δ_{aB} = allowable story drift for Structure B as defined in Table 12.12-1

h_{sx} = story height used in the definition of the allowable drift Δ_a in Table12.12-1. Note that Δ_d/ h_{sx} = the drift index.

The effects of seismic relative displacements shall be considered in combination with displacements caused by other loads as appropriate.

Analysis and Commentary

The seismic relative displacements to be used in the design of displacement-sensitive nonstructural components is D_pI_e instead of D_p, where D_p is given by Eqs. 13.3-6 to 13.3-9 and I_e is the Importance Factor set forth in Table 1.5-2. D_p in Eqs. 13.3-6 to 13.3-9 is based on deflection, δ_x, as given by Eq. (12.8-15), which, for the purposes of meeting the drift limit requirements of Table 12.12-1, correctly excludes the effect of the importance factor, I_e. For Risk Category III and IV structures, nonstructural components should be designed for the deflection that is amplified by I_e.

The changes to the definitions of the δ_x terms in Section 13.3.2.2 are necessary to ensure that the pointer is to δ_x in Section 12.8.6 and not to δ_{xe}.

Illustration of Relative Seismic Displacement

13.4.1, 13.4.2, 13.4.2.1, 13.4.2.2, 13.4.2.3

Modification

Design Force in Attachment and Anchors in Concrete or Masonry

At a Glance

The R_p penalty for anchors in concrete and masonry is revised. Standards are referenced for design of (1) anchors in concrete, (2) anchors in masonry, and (3) post-installed anchors in concrete and masonry.

2010 Standard

13.4.1 Design Forces~~.~~ in the Attachment

The force in the attachment shall be determined based on the prescribed forces and displacements for the component as determined ~~specified~~ in Sections 13.3.1 and 13.3.2~~.~~ except that R_p shall not be taken as larger than 6.

13.4.2 Anchors in Concrete or Masonry. ~~Anchors embedded in concrete or masonry shall be proportioned to carry the least of the following:~~

~~a. 1.3 times the force in the component and its supports due to the prescribed forces.~~

~~b. The maximum force that can be transferred to the anchor by the component and its supports.~~

~~The value of R_p used in Section 13.3.1 to determine the forces in the connected part shall not exceed 1.5 unless~~

~~a. The component anchorage is designed to be governed by the strength of a ductile steel element, or~~

~~b. The design of post-installed anchors in concrete used for the component anchorage is prequalified for seismic applications in accordance with ACI 355.2, or~~

~~c. The anchor is designed in accordance with Section 14.2.2.17.~~

13.4.2.1 Anchors in Concrete

Anchors in concrete shall be designed in accordance with Appendix D of ACI 318.

13.4.2.2 Anchors in Masonry

Anchors in masonry shall be designed in accordance with TMS 402/ACI 503/ASCE 5. Anchors shall be designed to be governed by the tensile or shear strength of a ductile steel element.

EXCEPTION: Anchors shall be permitted to be designed so that the attachment that the anchor is connecting to the structure undergoes ductile yielding at a load level corresponding to anchor forces not greater than their design strength, or the minimum design strength of the anchors shall be at least 2.5 times the factored forces transmitted by the component.

13.4.2.3 Post-installed Anchors in Concrete and Masonry

Post-installed anchors in concrete shall be pre-qualified for seismic applications in accordance with ACI 355.2 or other approved qualification procedures. Post-installed anchors in masonry shall be prequalified for seismic applications in accordance with approved qualification procedures.

Analysis and Commentary

The revisions address several issues associated with anchor design in Chapter 13:

1. There were a number of problems with the ASCE 7-05 provisions. The R_p penalty was intended to only apply to the anchorage; the language applied it to the entire component. The requirements for avoiding the R_p penalty were illogical. An anchor qualified in accordance with ACI 355.2 does not produce a ductile failure unless it is designed to do so. The list should have read: a) …or…b) …and…c). Finally, the reference to Section 14.2.2.17 raised the possibility that the 2.5 multiplier in that section for non-ductile anchors could be applied on top of the 1.3 factor and the R_p penalty in 13.4.2.
2. All post-installed anchors in concrete and masonry should be qualified for seismic loading through appropriate testing. The requisite tests for anchors in concrete are given in ACI 355.2 (or the ICC-ES criteria that incorporate these provisions: AC193, AC308); for post-installed anchors in masonry, the only standards for seismic prequalification are given in the ICC-ES acceptance criteria AC01, AC58, and AC106. All anchors in concrete should be designed in accordance with ACI 318 App. D. Anchors in masonry should be designed in accordance with ACI 530.
3. ASCE 7-05 Section 14.2.2.17 contained a ductility requirement and a force multiplier for nonductile anchors. These have now been incorporated into ACI 318-08 and are removed from ASCE 7-10 Chapter 14. The force multiplier (in ACI 318-08, it is a reduced φ-factor) replaces all other factors for non-ductile anchors in concrete.
4. The requirements for anchors in masonry are different than those for anchors in concrete and so must be treated separately. This is accomplished by creating new subparagraphs.
5. The limit on R_p (Section 13.4.1) is intended to address cases (e.g. high-deformability pipe) where the calculated connection forces could be artificially low, especially for welded or bolted connections where additional conservatism is not incorporated in the resistance calculation.

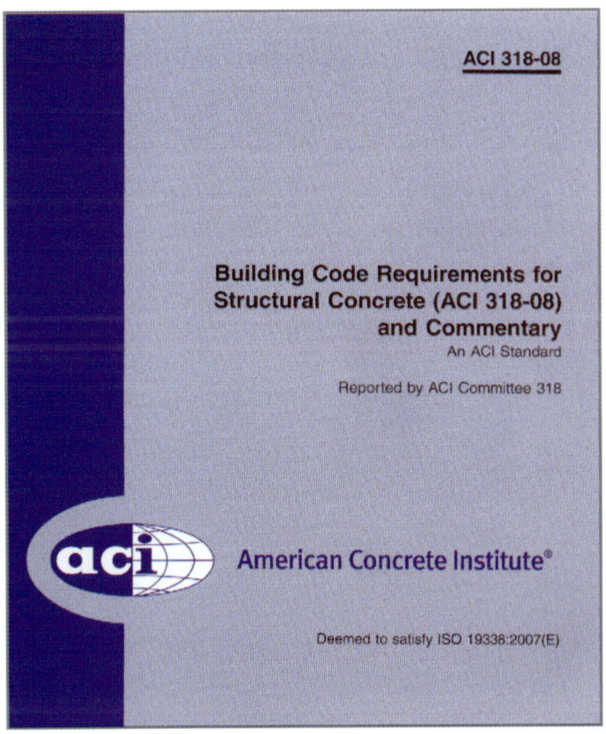

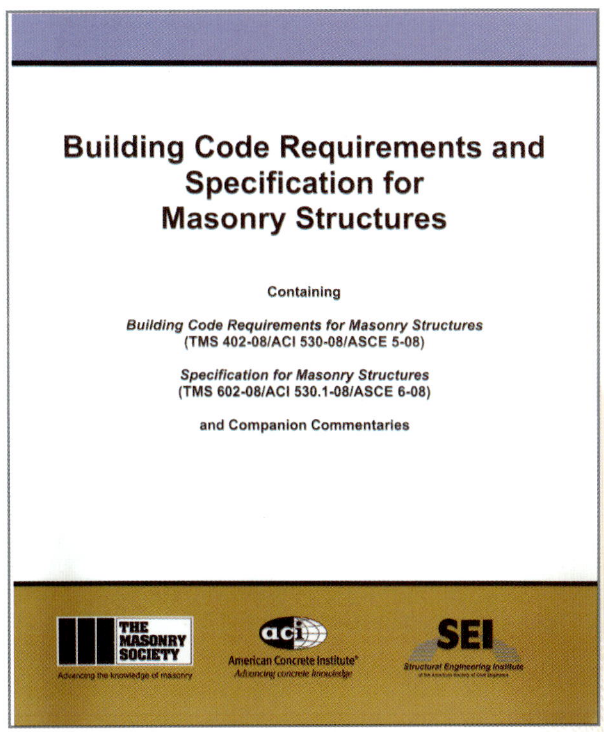

Referenced Standards for Design of Anchors in Concrete and Masonry

13.4.5

Power Actuated Fasteners

Modification and Addition

At a Glance

Power actuated fasteners in concrete and steel are now addressed separately and an exception is added to address lightly loaded power actuated fasteners in both concrete and steel.

2010 Standard

13.4.5 Power Actuated Fasteners

Power actuated fasteners in concrete or steel shall not be used for sustained tension loads or for brace applications in Seismic Design Categories D, E, or F unless approved for ~~such~~ seismic loading. Power actuated fasteners in masonry are not permitted unless approved for seismic loading.

EXCEPTION: Power actuated fasteners in concrete used for support of acoustical tile or lay-in panel suspended ceiling applications and distributed systems where the service load on any individual fastener does not exceed 90 lb (400 N). Power actuated fasteners in steel where the service load on any individual fastener does not exceed 250 lb (1,112 N).

Analysis and Commentary

Power actuated fasteners are used in both concrete and steel. The change recognizes this and sets forth separate restrictions for their use in each case.

This section was originally intended to address documented failures in the 1994 Northridge earthquake of power actuated fasteners used to hang sprinkler systems. While most of the failures involved the use of power actuated fasteners in concrete, some were also in steel. The degree to which the failures occurred because of poor installation, product deficiency, overload, or consequential damage is unclear from the record. Generally speaking, power actuated fasteners exhibit far better performance (mean, ultimate, and COV) in steel than in concrete. This is reflected in the provision, which still allows the use of power actuated fasteners in concrete for light distributed systems (e.g. lay-in ceilings).

Currently, no accepted procedure exists for the qualification of power actuated fasteners to resist seismic loads.

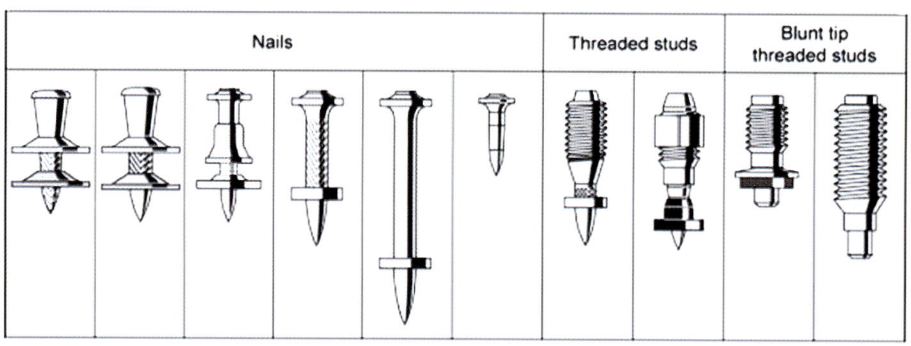

Power Actuated Fasteners

Table 13.5-1, Table 15.4-2

Modification

Signs and Billboards

At a Glance

The R and R_p values for signs and billboards are made consistent; both are equal to 3.0.

2010 Standard

Table 13.5-1 Coefficients for Architectural Components

Architectural Component	a_p[a]	R_p[b]
Signs and Billboards	2.5	3.0 ~~2.5~~

Table 15.4-2 Seismic Coefficients for Nonbuilding Structures not Similar to Buildings

Nonbuilding Structure Type	Detailing Requirements[c]	R	Ω_0	C_d	Structural Height, h_n, Limits (ft)[a, d]				
					A & B	C	D	E	F
Signs and Billboards		3.0 ~~3.5~~	1.75	3	NL	NL	NL	NL	NL

Analysis and Commentary

The values of R_p and R for "Signs and Billboards" specified in Chapters 13 and 15 should be consistent. This change provides for that consistency. The value being specified is the average of the values provided in ASCE 7-05 Chapters 13 and 15 and is considered to be a reasonable and acceptable value based on the collective judgment of the committees responsible for these chapters.

Signs and billboards need to be designed using R_p = 3 in accordance with Table 13.5-1 or R = 3 in accordance with Table 15.4-2.

13.5.6, 13.5.6.1, 13.5.6.2, 13.5.6.3, 13.5.8

Modification and Addition

Suspended Ceilings

At a Glance

ASTM E580 is referenced and modifications are made to provide consistency between referenced standards and ASCE 7-10.

2010 Standard

13.5.6 Suspended Ceilings

Suspended ceilings shall be in accordance with this section.

EXCEPTIONS:

1. Suspended ceilings with areas less than or equal to 144 ft^2 (13.4 m^2) that are surrounded by walls or soffits that are laterally braced to the structure above are exempt from the requirements of this section.
2. Suspended ceilings constructed of screw- or nail-attached gypsum board on one level that are surrounded by and connected to walls or soffits that are laterally braced to the structure above are exempt from the requirements of this section.

13.5.6.1 Seismic Forces

The weight of the ceiling, W_p, shall include the ceiling grid; ceiling tiles or ~~and~~ panels; light fixtures if attached to, clipped to, or laterally supported by the ceiling grid; and other components that are laterally supported by the ceiling. W_p shall be taken as not less than 4 psf (192 N/m^2).

The seismic force, F_p, shall be transmitted through the ceiling attachments to the building structural elements or the ceiling-structure boundary.

13.5.6.2 Industry Standard Construction for Acoustical Tile or Lay-In Panel Ceilings

Unless designed in accordance with Section 13.5.6.3, or seismically qualified in accordance with Section 13.2.5 or 13.2.6, ~~suspended~~ acoustical tile or lay-in panel ceilings shall be designed and constructed in accordance with this section.

13.5.6.2.1 Seismic Design Category C ~~Suspended~~ Acoustical tile or lay-in panel ceilings in structures assigned to Seismic Design Category C shall be designed and installed in accordance with ASTM C635, ASTM C636, and ~~the CISCA for Seismic Zones 0-2, except that seismic forces shall be determined in accordance with Sections 13.3.1 and 13.5.6.1.~~ ASTM E580, Section 4 - Seismic Design Category C.

13.5.6.2.2 Seismic Design Categories D through F ~~Suspended~~ Acoustical tile or lay-in panel ceilings in Seismic Design Categories D, E, and F shall be designed and installed in accordance with ASTM C635, ASTM C636, and ~~the CISCA for Seismic Zones 3-4~~ ASTM E580, Section 5 - Seismic Design Categories D, E, and F as modified by ~~the following~~ this section.

Acoustical tile or lay-in panel ceilings shall also comply with the following:

~~a. A heavy duty T-bar grid system shall be used.~~

a. ~~b.~~The width of the perimeter supporting closure angle or channel shall be not less than 2.0 in. (50 mm). Where perimeter supporting clips are used, they shall be qualified in accordance with approved test criteria. In each orthogonal horizontal direction, one end of the ceiling grid shall be attached to the closure angle or channel. The other end in each horizontal direction shall have a 0.75 in. (19 mm) clearance from the wall and shall rest upon and be free to slide on a closure angle or channel.

~~b. For ceiling areas exceeding 1,000 ft2 (92.9 m2), horizontal restraint of the ceiling to the structural system shall be provided. The tributary areas of the horizontal restraints shall be approximately equal.~~

> ~~**EXCEPTION:** Rigid braces are permitted to be used instead of diagonal splay wires. Braces and attachments to the structural system above shall be adequate to limit relative lateral deflections at point of attachment of ceiling grid to less than 0.25 in. (6 mm) for the loads prescribed in Section 13.3.1.~~

b. ~~d.~~ For ceiling areas exceeding 2,500 ft^2 (232 m^2), a seismic separation joint or full height partition that breaks the ceiling up into areas not exceeding 2,500 ft^2 (232 m^2), each with a ratio of the long to short dimension less than or equal to 4, shall be provided unless structural analyses are performed of the ceiling bracing system for the prescribed seismic forces that demonstrate ceiling ~~system~~ penetrations and closure angles or channels provide sufficient clearance to accommodate the anticipated lateral displacement. Each area shall be provided with closure angles or channels in accordance with ~~Item b~~ Section 13.5.6.2.2.a and horizontal restraints or bracing ~~in accordance with item e~~.

~~e. Except where rigid braces are used to limit lateral deflections, sprinkler heads and other similar type penetrations that do not behave integrally with the ceiling system in the lateral direction shall have a 2 in. (50 mm) oversize ring, sleeve, or adapter through the ceiling tile to allow for free movement of at least 1 in. (25 mm) in all horizontal directions. Alternatively, a swing joint or flexible device that can accommodate 1 in. (25 mm) of ceiling movement in all horizontal directions is permitted to be provided at the top of the sprinkler head extension.~~

~~f. Changes in ceiling plan elevation shall be provided with positive bracing.~~

~~g. Cable trays and electrical conduits shall be supported independently of the ceiling.~~

~~h. Suspended ceilings shall be subject to the special inspection requirements of Section 11A.1.3.9 of this standard.~~

13.5.6.3 Integral Construction

As an alternate to providing large clearances around sprinkler system penetrations through ceilings ~~systems~~, the sprinkler system and ceiling grid are permitted to be designed and tied together as an integral unit. Such a design shall consider the mass and flexibility of all elements involved, including the ceiling ~~system~~, sprinkler system, light fixtures, and mechanical (HVAC) appurtenances. Such design shall be performed by a registered design professional.

13.5.8 Partitions

13.5.8.1 General

Partitions that are tied to the ceiling and all partitions greater than 6 ft (1.8 m) in height shall be laterally braced to the building structure. Such bracing shall be independent of any ceiling ~~splay~~ lateral force bracing. Bracing shall be spaced to limit horizontal deflection at the partition head to be compatible with ceiling deflection requirements as determined in Section 13.5.6 for suspended ceilings and elsewhere in this section for other systems.

EXCEPTION: *(no changes to exception)*

Analysis and Commentary

Changes have been made to clarify and expand the requirements for suspended ceiling systems. Some terminology is changed to be consistent with terminology in referenced standards. The following is a summary of these changes:

1. *"Suspended ceiling"* is changed to *"acoustical tile or lay-in panel ceiling"* in Sections 13.5.6.2, 13.5.6.2.1, and 13.5.6.2.2. Since any ceiling suspended from the building structure is considered a "suspended ceiling", language was used from the referenced standards to communicate the scope of coverage. This was especially needed due to the exclusion of drywall ceilings, which are typically suspended.
2. *"Closure angle"* is changed to *"closure angle or channel"* because channels are occasionally used.
3. *"Splay bracing"* is changed to *"lateral force bracing."* There has been continuing confusion about what constitutes the lateral force bracing. Using several terms just exacerbated this. For less than 1,000 sq. ft, there is often a question as to whether the compression posts are waived along with the splay braces (they are).
4. *"Ceiling system"* is changed to *"ceiling."*

Two exceptions for suspended ceilings minor in nature are added, one for suspended ceilings less than 144 sq. ft in area and one for flat, drywall ceilings. These exceptions were in the referenced standards; however, by placing them at the beginning of Section 13.5.6, they will be more visible to the code user and applicable to all suspended ceilings. A reference to the option for seismic qualification of acoustical tile or lay-in panel ceilings that do not meet the prescriptive requirements of the section is added. ASTM E580, *Standard Practice for Installation of Ceiling Suspension Systems for Acoustical Tile and Lay-in Panels in Areas Subject to Earthquake Ground Motions*, is now referenced instead of CISCA standards, and much of the language that has been deleted from the ASCE 7-05 provisions is found in ASTM E580. This change simplifies the section, as most requirements are found in ASTM E580, instead of being split between two different CISCA documents and ASCE 7. A reference to the substitution of perimeter clips for the 2-in. wall angle is added as these are in common use and have been qualified using ICC-ES AC156 by multiple manufacturers.

Photo from "Seismic Fragility of Suspended Ceiling Systems", Technical Report MCEER-06-001
Hiram Badillo-Almaraz, Andrew S. Whittaker, Andrei M. Reinhorn and Gian Paolo Cimellaro.

Table 13.6-1 Modification

Seismic Coefficients for Electrical Conduit, Bus Ducts, Cable Trays, and Plumbing

At a Glance

Revisions are made to the seismic coefficients for electrical conduits and cable trays

2010 Standard

Table 13.6-1 Seismic Coefficients for Mechanical and Electrical Components

Mechanical and Electrical Components *(no changes to table entries)*	$a_p{}^a$	$R_p{}^b$
Vibration Isolated Components and Systems [b] *(no changes to table entries)*		
Distribution Systems		
Piping in accordance with ASME B31, including in-line components with joints made by welding or brazing.	2.5	12.0
Piping in accordance with ASME B31, including in-line components, constructed of high or limited deformability materials, with joints made by threading, bonding, compression couplings, or grooved couplings.	2.5	6.0
Piping and tubing not in accordance with ASME B31, including in-line components, constructed of high deformability materials, with joints made by welding or brazing.	2.5	9.0
Piping and tubing not in accordance with ASME B31, including in-line components, constructed of high or limited deformability materials, with joints made by threading, bonding, compression couplings, or grooved couplings.	2.5	4.5
Piping and tubing constructed of low deformability materials, such as cast iron, glass, and nonductile plastics.	2.5	3.0
Ductwork, including in-line components, constructed of high-deformability materials, with joints made by welding or brazing.	2.5	9.0
Ductwork, including in-line components, constructed of high- or limited-deformability materials with joints made by means other than welding or brazing.	2.5	6.0
Ductwork, including in-line components, constructed of low-deformability materials, such as cast iron, glass, and nonductile plastics.	2.5	3.0

Electrical conduit~~, bus ducts, rigidly mounted~~ and cable trays~~, and plumbing.~~	2.5 ~~1.0~~	6.0 ~~2.5~~
Bus ducts	1.0	2.5
Plumbing	1.0	2.5
Manufacturing or process conveyors (nonpersonnel)	2.5	3.0
~~Suspended cable trays~~	~~2.5~~	~~6.0~~

(No changes to footnotes)

Analysis and Commentary

The ASCE 7-05 entry "electrical conduit, bus ducts, rigidly mounted cable trays and plumbing" is revised to "electrical conduit and cable trays" and "bus ducts" and "plumbing" are made into separate entries. Conduit and cable trays are typically not rigid components, as implied by the ASCE 7-05 assignment of $a_p = 1.0$. For distributed electrical systems, the ASCE 7-05 values of $a_p = 1.0$ and $R_p = 2.5$ yield an a_p/R_p ratio of 0.40. With the change, $a_p/R_p = 2.5/6.0 = 0.42$, which yields a similar design value, but recognizes the inherent flexibility of the distributed electrical systems. The distinction between suspended and rigidly-mounted cable trays is deleted. Values for bus ducts are left unchanged, but are placed on a different line, since these do tend to behave as rigid elements. Values for plumbing are left unchanged, but are placed on a different line.

	a_p	R_p
Electrical conduit, bus ducts, rigidly mounted cable trays, and plumbing.	1.0	2.5

	a_p	R_p
Electrical conduit and cable trays	2.5	6.0
Bus Ducts	1.0	2.5
Plumbing	1.0	2.5

Table 13.5-1 and Table 13.6-1

Deletion

Chimneys and Stacks

At a Glance

Stacks are deleted from Table 13.5-1 and chimneys are deleted from Table 13.6-1.

2010 Standard

Table 13.5-1 Coefficients for Architectural Components

Architectural Component	$a_p{}^a$	$R_p{}^b$
Cantilever Elements (Unbraced or braced to structural frame below its center of mass)		
Parapets and cantilever interior nonstructural walls	2.5	2.5
Chimneys ~~and stacks~~ where laterally braced or supported by the structural frame	2.5	2.5
Cantilever Elements (Braced to structural frame above its center of mass)		
Parapets	1.0	2.5
Chimneys ~~and Stacks~~	1.0	2.5
Exterior Nonstructural Walls[b]	1.0[b]	2.5

Table 13.6-1 Seismic Coefficients for Mechanical and Electrical Components

Mechanical and Electrical Components	$a_p{}^a$	$R_p{}^b$
Roof-mounted ~~chimneys,~~ stacks, cooling and electrical towers laterally braced below their center of mass.	2.5	3.0
Roof-mounted ~~chimneys,~~ stacks, cooling and electrical towers laterally braced above their center of mass.	1.0	2.5

Analysis and Commentary

Chimneys and stacks were listed in ASCE 7-05 Table 13.5-1 as architectural components as well as in ASCE 7-05 Table 13.6-1 as mechanical and electrical components. To clarify this situation, "stacks" is removed from Table 13.5-1, so that it is covered only by Table 13.6-1, as almost all stacks are industrial in nature and are considered to be part of the mechanical exhaust system. Likewise, "chimneys" is removed from Table 13.6-1, because roof-mounted chimneys are in reality architectural components.

Chimneys are considered architectural components, and stacks are considered mechanical components for the purpose of Chapter 13.
Courtesy: Prabuddha Dasgupta

13.6.5.5 and 13.6.5.6

Modification and Addition

Additional Requirements for Mechanical and Electrical Components

At a Glance

The provisions in Section 13.6.5.5 are modified, improved, and clarified and a new Section 13.6.5.6 is added specifically for raceways.

2010 Standard

13.6.5.5 Additional Requirements

The following additional requirements shall apply to mechanical and electrical component supports:

1. Seismic supports shall be constructed so that support engagement is maintained.
2. ~~Oversized plate washers or other r~~Reinforcement (e.g. stiffeners or Belleville washers) shall be provided at bolted connections through ~~a~~ sheet metal equipment housings ~~base if the base is not reinforced with stiffeners or is not capable of transferring the required loads.~~ as required to transfer the equipment seismic loads specified in this section from the equipment to the structure. Where equipment has been certified per Section 13.2.2, 13.2.5, or 13.2.6, anchor bolts or other fasteners and associated hardware as included in the certification shall be installed in conformance with the manufacturer's instructions. For those cases where no certification exists or where instructions for such reinforcement are not provided, reinforcement methods shall be as specified by a registered design professional or as approved by the authority having jurisdiction.
3. *(no change to Item 3)*
4. *(no change to Item 4)*
5. ~~Expansion anchor shall not be~~ Where post-installed mechanical anchors are used for non-vibration isolated mechanical equipment rated over 10 hp (7.45 kW), they shall be qualified in accordance with ACI 355.2.

   ~~**EXCEPTION:** Undercut expansion anchors are permitted.~~

~~3. The supports for electrical distribution components shall be designed for the seismic forces and relative displacements defined in Sections 13.3.1 and 13.3.2 if any of the following conditions apply:~~

   ~~a. I_p is equal to 1.5 and conduit diameter is greater than 2.5 in. (64 mm) trade size.~~

   ~~b. Trapeze assemblies supporting conduit, and bus ducts or cable trays where I_p is equal to 1.5 and the total weight of the bus duct, cable tray, or conduit supported by trapeze assemblies exceeds 10 lb/ft (146 N/m).~~

   ~~c. Supports are cantilevered up from the floor.~~

   ~~d. Supports include bracing to limit deflection.~~

   ~~e. Supports are constructed as rigid welded frames.~~

   ~~f. Attachments into concrete utilize nonexpanding insets, power actuated fasteners, or cast iron embedments.~~

   ~~g. Attachments utilize spot welds, plug welds, or minimum size welds as defined by AISC.~~

6.~~7.~~ For piping, boilers, and pressure vessels, attachments to concrete shall be suitable for cyclic loads.

7.~~8.~~ For mechanical equipment, drilled and grouted-in-place anchors for tensile load applications shall use either expansive cement or expansive epoxy grout.

13.6.5.6 Conduit, Cable Tray, and Other Electrical Distribution Systems (Raceways)

Raceways shall be designed for seismic forces and seismic relative displacements as required in Section 13.3. Conduit greater than 2.5 inches (64 mm) trade size and attached to panels, cabinets, or other equipment subject to seismic relative displacement, D_p, shall be provided with flexible connections or designed for seismic forces and seismic relative displacements as required in Section 13.3.

EXCEPTIONS:

1. Design for the seismic forces and relative displacements of Section 13.3 shall not be required for raceways where either:
   a. Trapeze assemblies are used to support raceways and the total weight of the raceway supported by trapeze assemblies is less than 10 lb/ft (146 N/m), or
   b. The raceway is supported by hangers and each hanger in the raceway run is 12 in. (305 mm) or less in length from the raceway support point to the supporting structure. Where rod hangers are used, they shall be equipped with swivels to prevent inelastic bending in the rod.
2. Design for the seismic forces and relative displacements of Section 13.3 shall not be required for conduit, regardless of the value of I_p, where the conduit is less than 2.5 in. (64 mm) trade size.

Analysis and Commentary

The provisions in Section 13.6.5.5 are modified, improved, and clarified and a new Section 13.6.5.6 is added specifically for raceways.

Item 2 to Section 13.6.5.5 is revised to address additional requirements for reinforcement at bolted connections through sheet metal equipment housing. ASCE 7-05 language implied that an oversized plate washer was sufficient, regardless of the design of the sheet metal base. More recent proprietary testing conducted on equipment in accordance with Section 13.2.2 (Special Certification Requirements for Designated Seismic Systems) indicates that oversized plate washers may not be adequate to ensure good performance. Belleville washers offer one means of effectively transferring the inertial forces from the equipment to the structure. The use of Belleville washers improves the seismic performance of sheet metal connections by distributing the stress over a larger surface area of the sheet metal connection interface, allowing for bolted connections to be torqued to recommended values for proper pre-load while reducing the tendency for weak axis bending. The intrinsic spring loading capacity of the Belleville washer assists with long-term pre-load retention to maintain integrity of the seismic anchorage. Belleville washers are not proprietary and are widely available.

Item 5 is revised to eliminate the prohibition of expansion anchors when used for non-vibration-isolated mechanical equipment rated over 10 hp and instead allows them to be used if qualified in accordance with ACI 355.2. Many expansion anchors on the market today are suitable to resist vibratory loads. The qualification procedures of ACI 355.2 include tests for cyclic loading. The exception for undercut anchors is no longer required.

Item 6 is deleted and its requirements are revised and relocated to Section 13.6.5.6. This change accomplishes the following:

1. It consolidates the requirements for raceways (which include all forms of electrical distribution) in a separate section. The title of the section is specific as to what electrical components are being addressed. The term raceway is defined in several standards with somewhat varying language. As used here it is intended to describe all electrical distribution systems including conduit, cable trays, open and closed raceways, etc.

2. It requires design for seismic forces in accordance with Section 13.3 in general and provides exceptions where this is not required. Note that ASCE 7-05 language required design only where specific conditions were met. For example, ASCE 7-05 required that if rigid welded frames were used, then the supports had to be designed, but that if bolted frames were used, they did not have to be; also, if drilled and grouted-in-place anchors using expansive cement, etc. were used, the supports had to be designed, otherwise not. The responsible committee could not see any reason to preserve these requirements.
3. The new section simplifies the language to state that all supports must be designed, with two exceptions based on size and weight of the supported raceway. Experience indicates that a size limit of 2.5 inches can be established for the provision of flexible connections to accommodate seismic relative displacements that might occur between pieces of connected equipment, since smaller conduit normally possesses the required flexibility to accommodate such displacements. Where rod hangers are less than 12 inches in length, they are exempted from design only if they will not experience bending moments, i.e., by the provision of a swivel head at the top of the rod. Where this is not done and where braces are not provided, the rod hangers (and, where applicable, the anchors) must be designed for the resultant bending moments.

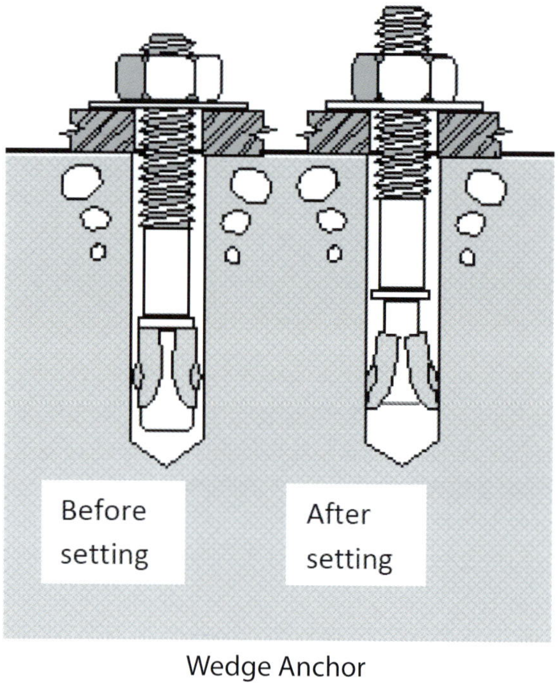

Example of
Torque-Controlled Expansion Anchor

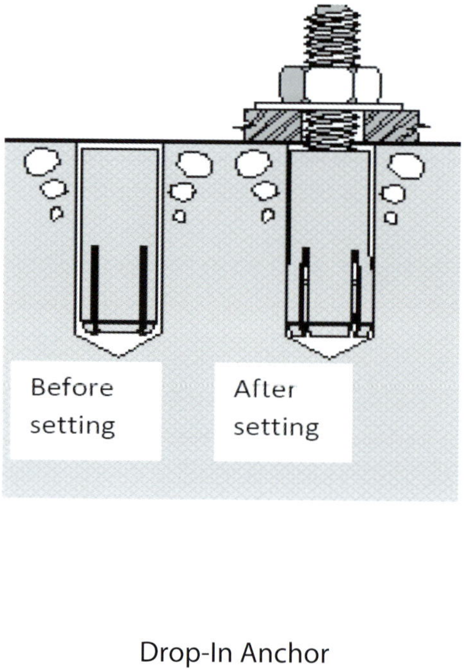

Example of
Torque-Controlled Expansion Anchor

13.6.7 Modification

Ductwork

At a Glance

Language in Section 13.6.7 is modified to be consistent with requirements found in standards addressing seismic bracing of ductwork.

2010 Standard

13.6.7 ~~HVAC~~ Ductwork. ~~Seismic supports are not required for HVAC ductwork with $I_p = 1.0$ if either of the following conditions are met for the full length of each duct run:~~

~~a. HVAC ducts are suspended from hangers 12 in. (305 mm) or less in length. The hangers shall be detailed to avoid significant bending of the hangers and their attachments, or~~

~~b. HVAC ducts have a cross-sectional area of less than 6 ft^2 (0.557 m^2).~~ HVAC and other ductwork shall be designed for seismic forces and seismic relative displacements as required in Section 13.3. Design for the displacements across seismic joints shall be required for ductwork with $I_p = 1.5$ without consideration of the exceptions below.

EXCEPTIONS:The following exceptions pertain to ductwork not designed to carry toxic, highly toxic, or flammable gases or used for smoke control:

1. Design for the seismic forces and relative displacements of Section 13.3 shall not be required for ductwork where either:
   a. Trapeze assemblies are used to support ductwork and the total weight of the ductwork supported by trapeze assemblies is less than 10 lb/ft (146 N/m); or
   b. The ductwork is supported by hangers and each hanger in the duct run is 12 in. (305 mm) or less in length from the duct support point to the supporting structure. Where rod hangers are used, they shall be equipped with swivels to prevent inelastic bending in the rod.

2. Design for the seismic forces and relative displacements of Section 13.3 shall not be required where provisions are made to avoid impact with larger ducts or mechanical components or to protect the ducts in the event of such impact; and HVAC ducts have a cross-sectional area of less than 6 ft^2 (0.557 m^2) or weigh 17 lb/ft (248 N/m) or less.

HVAC duct systems fabricated and installed in accordance with standards approved by the authority having jurisdiction shall be deemed to meet the lateral bracing requirements of this section.

Components that are installed in-line with the duct system and have an operating weight greater than 75 lb (334 N), such as fans, heat exchangers, and humidifiers, shall be supported and laterally braced independent of the duct system and such braces shall meet the force requirements of Section 13.3.1. Appurtenances such as dampers, louvers, and diffusers shall be positively attached with mechanical fasteners. Unbraced piping attached to in-line equipment shall be provided with adequate flexibility to accommodate ~~differential displacements~~ the seismic relative displacements of Section 13.3.2.

Analysis and Commentary

The new language clarifies that ductwork shall be designed to meet the requirements of Section 13.3 with exceptions. The language is consistent with revised language for other distribution systems (electrical and piping). Seismic bracing manuals developed by SMACNA (Sheet Metal and Air Conditioning Contractors National Association) and component suppliers have been used successfully in California and elsewhere. The new language is consistent with current limitations in those manuals.

The exceptions do not pertain to ductwork designed to carry toxic, highly toxic, or flammable gases or used for smoke control in order to better assure that ductwork subjected to lateral loads does not deflect excessively and leak, which would place occupants at risk.

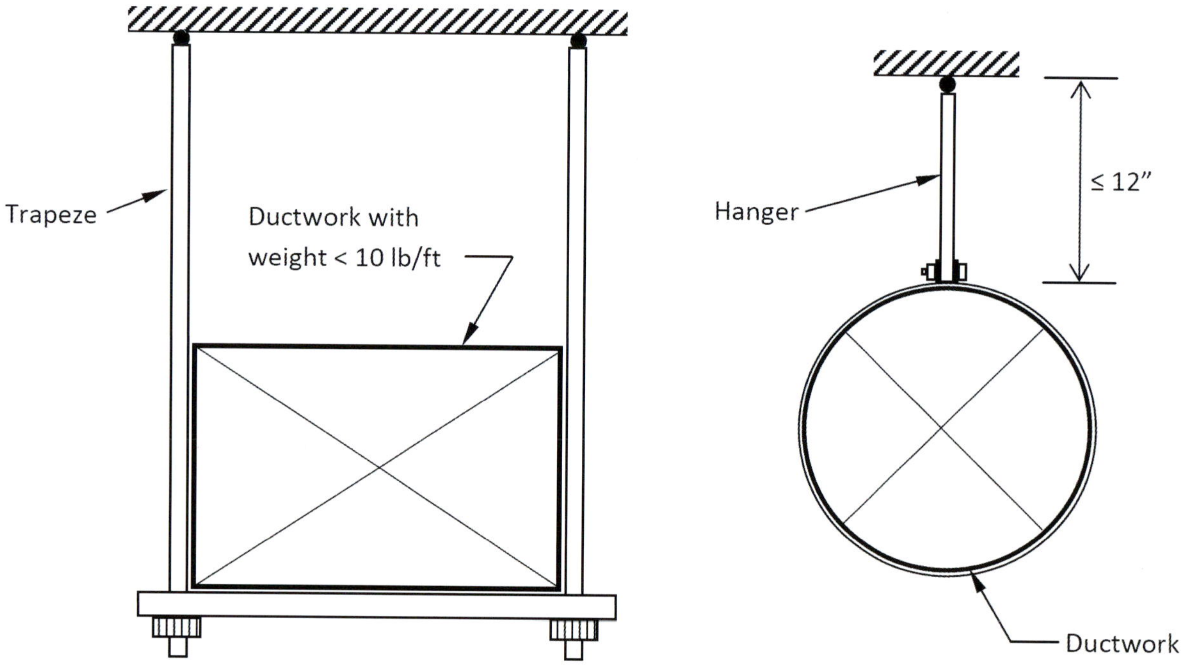

Illustration of Section 13.6.7, Exception 1

13.6.8, 13.6.8.1, 13.6.8.2, 13.6.8.3, 13.6.9, 13.6.10, 13.6.11

Modification

Piping Systems

At a Glance

Section 13.6.8 and its subsections are reorganized and revised to clarify the intent.

2010 Standard

13.6.8 Piping Systems

Unless otherwise noted in this section, ~~P~~piping systems shall be designed for the seismic forces and seismic relative displacements of Section 13.3. ~~satisfy the requirements of this section except that~~ ASME pressure piping systems shall satisfy the requirements of Section 13.6.8.1. Fire protection sprinkler piping shall satisfy the requirements of Section 13.6.8.2. elevator system piping shall satisfy the requirements of Section 13.6.10.

Where other applicable material standards or recognized design bases are not used, piping design including consideration of service loads shall be based on the following allowable stresses:

a. For piping constructed with ductile materials (e.g., steel, aluminum, or copper), 90 percent of the minimum specified yield strength.

b. For threaded connections in piping constructed with ductile materials, 70 percent of the minimum specified yield strength.

c. For piping constructed with nonductile materials (e.g., cast iron, or ceramics), 10 percent of the material minimum specified tensile strength.

d. For threaded connections in piping constructed with nonductile materials, 8 percent of the material minimum specified tensile strength.

Piping not detailed to accommodate the seismic relative displacements at connections to other components shall be provided with connections having sufficient flexibility to avoid failure of the connection between the components.

~~Except for piping designed and constructed in accordance with NFPA 13, seismic supports shall not be required for other piping systems where one of the following conditions is met:~~

1. ~~Piping is supported by rod hangers; hangers in the pipe run are 12 inches (305 mm) or less in length from the top of the pipe to the supporting structure; hangers are detailed to avoid bending of the hangers and their attachments; and provisions are made for piping to accommodate expected deflections.~~

2. ~~High deformability piping is used; provisions are made to avoid impact with larger piping or mechanical components or to protect the piping in the event of such impact; and the following size requirements are satisfied:~~

~~a. For Seismic Design Categories D, E or F where I_p is greater than 1.0, the nominal pipe size shall be 1 inch (25 mm) or less.~~

~~b. For Seismic Design Category C, where I_p is greater than 1.0,, the nominal pipe size shall be 2 inches (51 mm) or less.~~

~~c. For Seismic Design Category D, E or F where I_p is equal to 1.0, the nominal pipe size shall be 3 inches (76 mm) or less.~~

13.6.8.1 ASME Pressure Piping Systems

Pressure piping systems, including their supports, designed and constructed in accordance with ASME B31 shall be deemed to meet the force, displacement, and other requirements of this section. In lieu of specific force and displacement requirements provided in ASME B31, the force and displacement requirements of Sections 13.3~~.1 and 13.3.2~~ shall be used. Materials meeting the toughness requirements of ASME B31 shall be considered high-deformability materials.

13.6.8.2 Fire Protection Sprinkler Piping Systems ~~**in Seismic Design Category C.**~~

~~In structures assigned to Seismic Design Category C, f~~Fire protection sprinkler piping, pipe hangers, and bracing ~~systems~~ designed and constructed in accordance with NFPA 13 shall be deemed to meet the force and displacement ~~other~~ requirements of this section. The exceptions of Section 13.6.8.3 shall not apply.

13.6.8.3 Exceptions

Design of piping systems and attachments for the seismic forces and relative displacements of Section 13.3 shall not be required where one of the following conditions apply:

1. Trapeze assemblies are used to support piping whereby no single pipe exceeds the limits set forth in 3a, 3b, or 3c below and the total weight of the piping supported by the trapeze assemblies is less than 10 lb/ft (146 N/m).

2. The piping is supported by hangers and each hanger in the piping run is 12 in. (305 mm) or less in length from the top of the pipe to the supporting structure. Where pipes are supported on a trapeze, the trapeze shall be supported by hangers having a length of 12 in. (305 mm) or less. Where rod hangers are used, they shall be equipped with swivels, eye nuts, or other devices to prevent bending in the rod.

3. Piping having an R_p in Table 13.6-1 of 4.5 or greater is used and provisions are made to avoid impact with other structural or nonstructural components or to protect the piping in the event of such impact and where the following size requirements are satisfied:
   a. For Seismic Design Category C where I_p is greater than 1.0, the nominal pipe size shall be 2 in. (50 mm) or less.
   b. For Seismic Design Categories D, E, or F and values of I_p are greater than 1.0, the nominal pipe size shall be 1 in. (25 mm) or less.
   c. For Seismic Design Categories D, E, or F where $I_p = 1.0$, the nominal pipe size shall be 3 in. (80 mm) or less.

~~**13.6.8.3 Fire Protection Sprinkler Systems in Seismic Design Categories D through F.** In structures assigned to Seismic Design Category D, E or F, the following requirements shall be satisfied:~~

~~1. The hangers and sway bracing of the fire protection sprinkler systems shall be deemed to meet the requirements of this section if both of the following requirements are satisfied:~~

   ~~a. The hangers and sway bracing are designed and constructed in accordance with NFPA 13.~~
   ~~b. The force and displacement requirements of Sections 13.3.1 and 13.3.2 are satisfied.~~
~~2. The fire protection sprinkler system piping itself shall meet the force and displacement requirements of Section 13.3.1 and13.3.2.~~
~~3. The design strength of the fire protection sprinkler system piping for seismic loads in combination with other service loads and appropriate environmental effects shall be based on the following material properties:~~
   ~~a. For piping and components constructed with ductile materials (e.g., steel, aluminum, or copper), 90 percent of the minimum specified yield strength.~~
   ~~b. For threaded connections in components constructed with ductile materials, 70 percent of the minimum specified yield strength.~~
   ~~c. For piping and components constructed with nonductile materials (e.g., plastic, cast iron, or ceramics), 10 percent of the material minimum specified tensile strength.~~

13.6.9 Boilers and Pressure Vessels

Boilers and pressure vessels designed and constructed in accordance with ASME BPVC shall be deemed to meet the force, displacement, and other requirements of this section. In lieu of specific force and displacement requirements provided in ASME BPVC, the force and displacement requirements of Sections 13.3.1 and 13.3.2 shall be used. Materials meeting the toughness requirements of ASME BPVC shall be considered high-deformability materials. Other boilers and pressure vessels designated as having an $I_p = 1.5$, but not designed and constructed in accordance with the requirements of ASME BPVC, shall comply with the requirements of Section 13.6.11.

13.6.10 Elevator and Escalator Design Requirements

Elevators and escalators designed in accordance with the seismic requirements of ASME A17.1 shall be deemed to meet the seismic force requirements of this section, except as modified in the following text. The exceptions of Section 13.6.8.3 shall not apply to elevator piping.

13.6.11 Other Mechanical and Electrical Components

Mechanical and electrical components, including conveyor systems, not designed and constructed in accordance with the reference documents in Chapter 23 shall meet the following:

1. *(No change in Item 1)*
2. For mechanical components with hazardous substances and assigned a component importance factor, I_p, of 1.5 in accordance with Section 13.1.3, and for boilers and pressure vessels not designed in accordance with ASME BPVC, the design strength for seismic loads in combination with other service loads and appropriate environmental effects shall be based on the following material properties:
   (no change in Items a. through c.)
   d. For threaded connections in ~~piping~~ components constructed with nonductile materials, 8 percent of the material minimum specified tensile strength.

Analysis and Commentary

Section 13.6.8 and its subsections are reorganized to clarify the intent by first stating requirements for common piping with default stress values appropriate for the type of piping and setting forth separate requirements for ASME, NFPA, and elevator piping. Section 13.6.11 (Other Mechanical and Electrical Components) is revised to address components other than piping. The allowable stresses set forth in Section 13.6.8, a. and b. are based on values consistent with structural steel standards for comparable piping materials. The allowable stresses set forth in Section 13.6.8, c. and d. are derived from values consistent with ASME standards for comparable piping materials.

The detailed provisions for fire sprinkler systems are deleted in view of the overhaul of the requirements of NFPA 13 in the 2007 edition. Maximum load requirements have been implemented for the lateral sway braces in NFPA 13 along with maintaining the already-existing maximum spacing requirements.

The question has arisen several times regarding whether ASME constructions are high-deformability or otherwise. The ASME piping and pressure vessel documents provide toughness requirements to be met, which assure that ASME constructions will provide adequate deformability during operations, and a sentence is added to Section 13.6.8.1 indicating that materials meeting the toughness requirements of ASME B3 are to be considered high-deformability materials

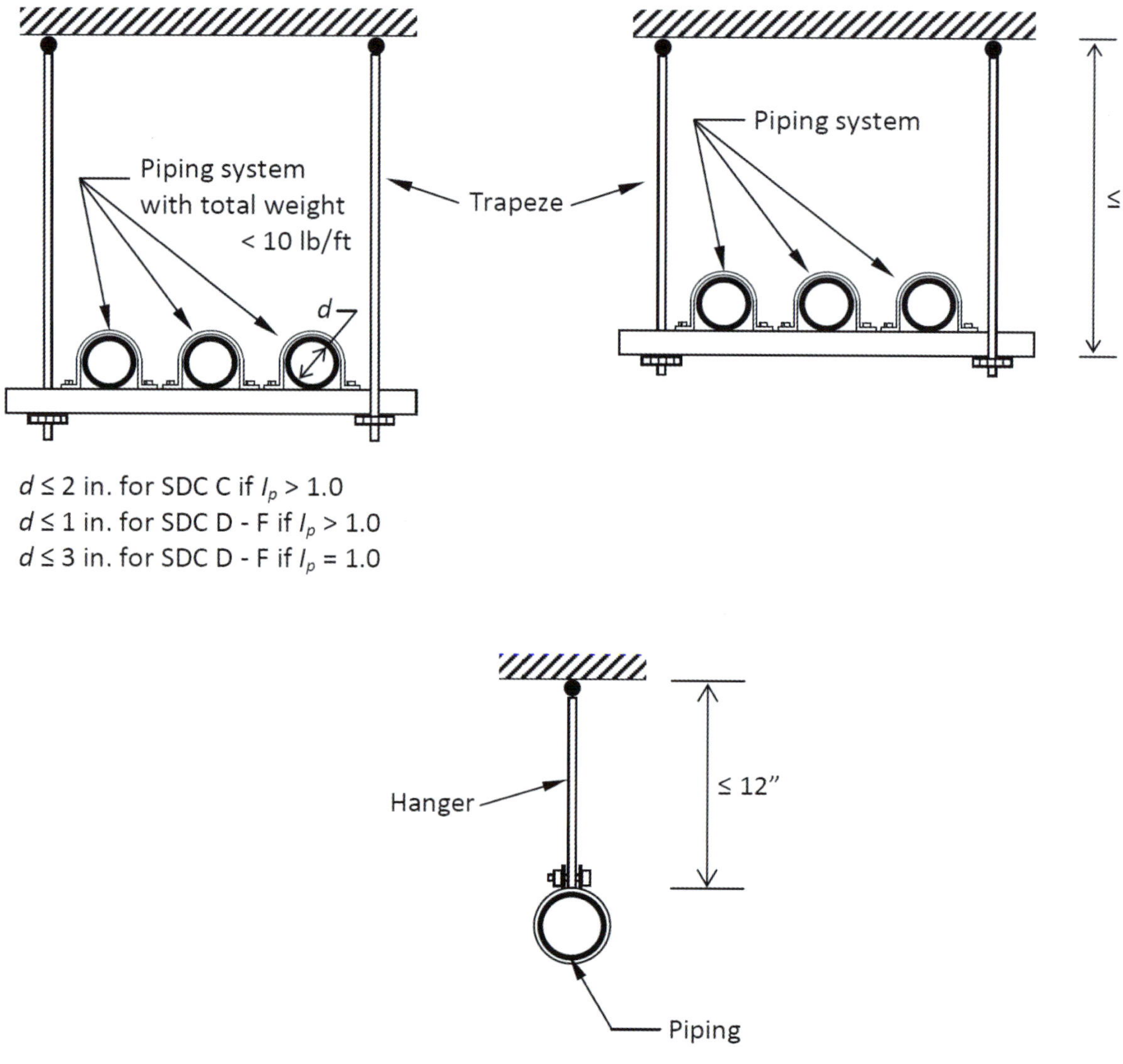

Illustration of Section 13.6.8.3 Items 1 and 2 (not applicable to NFPA 13 designs)

13.6.10.3 Modification

Seismic Controls for Elevators

At a Glance

Changes in Section 13.6.10.3 are made to achieve consistency with ASME A17.1, *Safety Code for Elevators and Escalators,* and an exception is added to eliminate unnecessary elevator shutdowns.

2010 Standard

13.6.10.3 Seismic ~~Switches~~ Controls for Elevators

Elevators operating with a speed of 150 ft/min (46 m/min) or greater shall be provided with seismic switches. Seismic switches shall provide an electric signal indicating that structural motions are of such a magnitude that the operation of the elevators may be impaired. ~~The seismic switch shall be located at or above the highest floor serviced by the elevators. The seismic switch shall have two horizontal perpendicular axes of sensitivity. Its trigger level shall be set to 30 percent of the acceleration of gravity.~~ Seismic switches in accordance with Section 8.4.10.1.2 of ASME A17.1 shall be deemed to meet the requirements of this section.

EXCEPTION: In cases where seismic switches cannot be located near a column in accordance with ASME A17.1, they shall have two horizontal axes of sensitivity and have a trigger level set to 20 percent of the acceleration of gravity where located at or near the base of the structure and 50 percent of the acceleration of gravity in all other locations.

Upon activation of the seismic switch, elevator operations shall conform to requirements of ASME A17.1, except as noted in the following text.

In facilities where the loss of the use of an elevator is a life-safety issue, the elevator shall only be used after the seismic switch has triggered provided that:

1. The elevator shall operate no faster than the service speed.
2. Before the elevator is occupied, it is operated from top to bottom and back to top to verify that it is operable.

Analysis and Commentary

Changes in Section 13.6.10.3 are made to achieve consistency with ASME A17.1, *Safety Code for Elevators and Escalators.* A17.1 requires mounting the seismic switch, when used exclusively to control elevators, in the machine room, and where possible, adjacent to a vertical load-bearing building structural member with its axis of sensitivity in the vertical direction and set to trigger at 0.15g. Placing the switch near a vertical structural support member will prevent significant amplification of vertical motions between the foundation and the seismic trigger.

At small and moderate ground motions, buildings often exhibit horizontal amplifications at the top of three or more. Thus, with horizontal ground motions of 0.1 g, a value commonly experienced in California, the ASCE 7-05 seismic switch could trigger elevator shutdowns. This required the elevator to be inspected by an elevator mechanic, adding unnecessary disruption and cost. The seismic switch used by A17.1 is designed to trigger on the P wave so that, in many cases, it will provide adequate time to stop the elevator and allow passengers to exit the elevator before severe shaking in the building starts. This earlier trigger is also more likely to allow passengers to exit the elevator prior to loss of building power, a common occurrence in moderate earthquakes. This early exit would avoid having passengers trapped in an elevator during an earthquake and the need for first responders to rescue them from the elevator. For cases where switches cannot be located near columns, i.e. where the vertical accelerations may not be transmitted to the switch, an exception to A17.1 is added to avoid unnecessary shutdowns. This exception was developed based on the procedures implemented in hospitals in California following the Northridge earthquake

2004 Edition of ASME A17.1

Part V
Chapter 15 Seismic Design Requirements for Nonbuilding Structures

15.4.1, Table 15.4-2, 15.6.2

Modification

Stacks and Chimneys

At a Glance

The seismic coefficients as well as structural system and height limits for reinforced and nonreinforced masonry structures not similar to buildings are changed in Table 15.4-2. A new entry is added to Table 15.4-2 for concrete chimneys and stacks designed in accordance with ACI 307 and new opening detailing requirements for this structure type are added to Section 15.6.2.

2010 Standard

15.4.1 Design Basis

(no change to text)

2.

(no change to text)

EXCEPTION: Tanks and vessels that are designed to AWWA D100, AWWA D103, API 650 Appendix E, and API 620 Appendix L as modified by this standard, and stacks and chimneys that are designed to ACI 307 as modified by this standard, shall be subject to the larger of the minimum base shear values defined . . .

(no changes to remainder of Exception)

Table 15.4-2 Seismic Coefficients for Nonbuilding Structures not Similar to Buildings

Nonbuilding Structure Type	Detailing Requirements[c]	R	Ω_0	C_d	Structural Height, h_n, Limits (ft)[a, d]				
					B	C	D	E	F
Cast-in-place concrete silos~~, stacks, and chimneys~~ having walls continuous to the foundation	15.6.2	3	1.75	3	NL	NL	NL	NL	NL
All other reinforced masonry structures not similar to buildings detailed as intermediate reinforced masonry shear walls	14.4.1[f]	3	2	2.5	NL	NL	~~NL~~50	50	50
All other reinforced masonry structures not similar to buildings detailed as ordinary reinforced masonry shear walls	14.4.1	2	2.5	1.75	NL	160	NP	NP	NP
All other nonreinforced masonry structures not similar to buildings	14.4.1	1.25	2	1.5	NL	NL	~~50~~NP	~~50~~NP	~~50~~NP
Concrete chimneys and stacks	15.6.2 and ACI 307	2	1.5	2.0	NL	NL	NL	NL	NL

15.6.2 Stacks and Chimneys

Stacks and chimneys are permitted to be either lined or unlined and shall be constructed from concrete, steel, or masonry. Steel stacks, concrete stacks, steel chimneys, concrete chimneys, and liners shall be designed to resist seismic lateral forces determined from a substantiated analysis using reference documents. Interaction of the stack or chimney with the liners shall be considered. A minimum separation shall be provided between the liner and chimney equal to C_d times the calculated differential lateral drift.

Concrete chimneys and stacks shall be designed in accordance with the requirements of ACI 307 except that (1) the design base shear shall be determined based on Section 15.4.1 of this standard; (2) the seismic coefficients shall be based the values provided in Table 15.4-2, and (3) openings shall be detailed as required below. When modal response spectrum analysis is used for design, the procedures of Section

12.9 shall be permitted to be used.

For concrete chimneys and stacks assigned to SDC D, E, and F, splices for vertical rebar shall be staggered such that no more than 50% of the bars are spliced at any section and alternate lap splices are staggered by the development length. In addition, where the loss of cross-sectional area is greater than 10%, cross sections in the regions of breachings/openings shall be designed and detailed for vertical force, shear force, and bending moment demands along the vertical direction, determined for the affected cross section using an overstrength factor of 1.5. The region where the overstrength factor applies shall extend above and below the opening(s) by a distance equal to half of the width of the largest opening in the affected region. Appropriate reinforcement development lengths shall be provided beyond the required region of overstrength. The jamb regions around each opening shall be detailed using the column tie requirements in Section 7.10.5 of ACI 318. Such detailing shall extend for a jamb width of a minimum of two times the wall thickness and for a height of the opening height plus twice the wall thickness above and below the opening, but no less than the development length of the longitudinal bars. Where the existence of a footing or base mat precludes the ability to achieve the extension distance below the opening and within the stack, the jamb reinforcing shall be extended and developed into the footing or base mat. The percentage of longitudinal reinforcement in jamb regions shall meet the requirements of Section 10.9 of ACI 318 for compression members.

Analysis and Commentary

The seismic coefficients as well as structural system and height limits for reinforced and nonreinforced masonry structures not similar to buildings are changed in Table 15.4-2, to make them consistent with the corresponding limits and restrictions in Chapter 12 and to add specific detailing requirements for reinforced masonry.

It was recognized that the ASCE 7-05 seismic requirements for stacks were very liberal (R =3) and that no detailing requirements at openings in reinforced concrete stacks were specified. Concrete chimneys have low ductility, and their seismic behavior is especially critical in the opening regions due to inherent reduction in strength and loss of confinement of vertical reinforcement in the jamb regions around the openings. Work was completed in consultation with ACI Committee 307, which concluded that an R value equal to 2 is more appropriate and detailing requirements around openings were developed. These requirements are consistent with new international standards for concrete stacks developed after the 1999 Turkey earthquake by CICIND (an international standard organization for industrial chimneys).

The revised provisions accomplish the following:

1. Stacks and chimneys designed to ACI 307 are added to the EXCEPTION list, which permits using the lower nonbuilding structure minimum base shear values. Stacks and chimneys designed to ACI 307 will have reliable performance similar to that of tanks and vessels designed to the listed standards and therefore should be allowed to use the lower minimum base shears.

2. Table 15.4-2 separates out stacks and chimneys designed to ACI 307 and the detailing requirements of Section 15.6.2 and provides values for R, Ω_0 and C_d. The R and Ω_0 values of 2.0 and 1.5, respectively, are based on the recommendations of the ACI 307 committee.

3. Section 15.6.2 is modified by providing charging language to permit use of ACI 307 for design under certain conditions, which will make the resulting design consistent with ASCE 7. It also provides language that permits the modal analysis scaling procedures of Section 12.9 to be used for design.

4. Language is added to Section 15.6.2 providing specific opening detailing requirements.

A new entry is added to Table 15.4-2 for concrete chimneys and stacks designed in accordance with ACI 307.

15.5.3, 23.1

Modification, Replacement

Steel Storage Racks

At a Glance

Steel storage racks supported at or below grade are now required to be designed in accordance with ANSI/RMI MH 16.1-2008, as modified in ASCE 7-10 Section 15.5.3. As an alternative, steel storage racks may continue to be designed by the provisions that were in ASCE 7-05.

2010 Standard

15.5.3 Steel Storage Racks

Steel storage racks supported at or below grade shall be designed in accordance with ANSI/RMI MH 16.1 and its force and displacement requirements, except as follows:

15.5.3.1

Modify Section 2.6.2 of ANSI/RMI MH 16.1 as follows:

2.6.2 Minimum Seismic Forces

The storage rack shall be designed...

Above-Grade Elevation: *Storage rack installed at elevations above grade shall be designed, fabricated and installed in accordance with the following requirements:*

Storage racks shall meet the force and displacement requirements required of nonbuilding structures supported by other structures, including the force and displacement effects caused by amplifications of upper-story motions. In no case shall the value of V be taken as less than the value of F_p determined in accordance with Section 13.3.1 of ASCE/SEI 7, where R_p is taken equal to R, and a_p is taken equal to 2.5.

15.5.3.2

Modify Section 7.2.2 of ANSI/RMI MH 16.1 as follows:

7.2.2 Base Plate Design

Once the required bearing area has been determined from the allowable bearing stress F'_p the minimum thickness of the base plate is determined by rational analysis or by appropriate test using a test load 1.5 times the ASD design load or the factored LRFD load. Design forces that include seismic loads for anchorage of steel storage racks to concrete or masonry shall be determined using load combinations with overstrength provided in Section 12.4.3.2 ASCE/SEI 7. The overstrength factor shall be taken as 2.0.

Anchorage of steel storage racks to concrete shall be in accordance with the requirements of Section 15.4.9 of ASCE/SEI 7. Upon request, information shall be given to the owner, or the owner's agent on the location, size, and pressures under the column base plates of each type of upright frame in the installation. When rational analysis is used to determine base plate thickness and other applicable standards do not apply, the base plate shall be permitted to be designed for the following loading conditions, where applicable: (balance of section unchanged)

15.5.3.3

Modify Section 7.2.4 of ANSI/RMI MH 16.1 as follows:

7.2.4 Shims

Shims may be used under the base plate to maintain the plumbness of the storage rack. The shims shall be made of a material that meets or exceeds the design bearing strength (LRFD) or allowable bearing strength (ASD) of the floor. The shim size and location under the base plate shall be equal to or greater than the required base plate size and location

~~*In no case shall the total thickness of any set of shims under a base plate exceed six times the diameter of the largest anchor bolt used in that base.*~~

Shims ~~that are a total thickness of less than or equal to six times the anchor bolt diameter under bases with less than two anchor bolts~~ shall be interlocked or welded together in a fashion that is capable of transferring all the shear forces at the base.

~~*Shims that are a total thickness of less than or equal to two times the anchor bolt diameter need not be interlocked or welded together.*~~

Bending in the anchor associated with shims or grout under the base plate shall be taken into account in the design of the anchor bolts.

15.5.3.4 Alternative

As an alternative to ANSI MH 16.1as modified above, steel storage racks shall be permitted to be designed in accordance with the requirements of Sections 15.1, 15.2, 15.3, 15.5.1, and 15.5.3.5 through 15.5.3.8 of this standard.

~~In addition to the requirements of Section 15.5.1, steel storage racks shall be designed in accordance with the requirements of Sections 15.5.3.1 through 15.5.3.4. Alternatively, steel storage racks are permitted to be designed in accordance with the method defined in Section 2.7 "Earthquake Forces" of RMI where the following changes are included:~~

1. ~~The values of C_a and C_v used shall equal $0.4S_{DS}$ and S_{D1}, respectively, where S_{DS} and S_{D1} are determined in accordance with Section 11.4.4 of this standard.~~
2. ~~The importance factor for storage racks in structures open to the public, such as retail warehouse stores, shall be taken equal to 1.5.~~
3. ~~For storage racks supported at or below grade, the value of C_S used shall not be less than $0.14S_{DS}$. For storage racks supported above grade, the value of C_S used shall not be less than the value for F_p determined in accordance with Section 13.3.1 of this standard where R_p is taken as equal to R from RMI and a_p is taken as equal to 2.5.~~

15.5.3.5~~1~~ *(no change from ASCE 7-05 Section 15.5.3.1)*

15.5.3.6~~2~~ *(no change from ASCE 7-05 Section 15.5.3.2)*

15.5.3.7~~3~~ *(no change from ASCE 7-05 Section 15.5.3.3)*

15.5.3.8~~4~~ *(no change from ASCE 7-05 Section 15.5.3.4)*

23.1 CONSENSUS STANDARDS AND OTHER REFERENCE DOCUMENTS

RMI
Rack Manufacturers Institute
8720 Red Oak Boulevard
Suite 201
Charlotte, NC 28217

ANSI/MH 16.1 ~~RMI~~
Section 15.5.3
Specification for the Design, Testing, and Utilization of Industrial Steel Storage Racks, 2008 ~~1997, reaffirmed 2002~~

Analysis and Commentary

ANSI MH 16.1-2008, *Specification for the Design, Testing, and Utilization of Industrial Steel Storage Racks,* is designed to be a stand-alone standard to be adopted by reference. For a number of years, the Rack Manufacturers Institute has worked to harmonize its *Specification* with the model codes, ASCE 7, and the NEHRP *Provisions*. Working with the Subcommittee on Nonbuilding Structures (TS-8) of the BSSC Provisions Update Committee (PUC), the Task Committee on Nonbuilding Structures (TC-8) of the Seismic Subcommittee of ASCE 7, and the Task Group on FEMA 460 (2005), *Seismic Considerations for Steel Storage Racks Located in Areas Open to the Public*, the seismic provisions have been developed and written to be self-sufficient, incorporating the results of those efforts.

The anchorage provisions of MH 16.1-2008 are not in conformance with those of ASCE 7. Therefore, anchorage provisions are provided as a modification to MH 16.1-2008. The same type of anchorage requirements are also added in a new Section 15.4.9 that provides requirements for nonbuilding structures.

Updated ANSI MH 16.1-2008 is referenced for storage racks in ASCE 7-10.

15.4.9, 15.7.3, 15.7.5, 15.7.11.7

Addition, Modification

Anchors in Concrete or Masonry

At a Glance

Section 15.4.9 is added to provide a consistent treatment of nonbuilding structures anchorage throughout Chapter 15. Coordinated changes are made in Section 15.7, Tanks and Vessels.

2010 Standard

15.4.9 Anchors in Concrete or Masonry

15.4.9.1 Anchors in Concrete

Anchors in concrete used for nonbuilding structure anchorage shall be designed in accordance with Appendix D of ACI 318.

15.4.9.2 Anchors in Masonry

Anchors in masonry used for nonbuilding structure anchorage shall be designed in accordance with TMS 402/ACI 530/ASCE 6. Anchors shall be designed to be governed by the tensile or shear strength of a ductile steel element.

EXCEPTION: Anchors shall be permitted to be designed so that the attachment that the anchor is connecting to the structure undergoes ductile yielding at a load level corresponding to anchor forces not greater than their design strength, or the minimum design strength of the anchors shall be at least 2.5 times the factored forces transmitted by the attachment.

15.4.9.3 Post-Installed Anchors in Concrete and Masonry

Post-installed anchors in concrete shall be prequalified for seismic applications in accordance with ACI 355.2 or other approved qualification procedures. Post-installed anchors in masonry shall be prequalified for seismic applications in accordance with approved qualification procedures.

15.7.3 Strength and Ductility

Structural ~~components and~~ members that are part of the seismic force-resisting ~~lateral support~~ system shall be designed to provide the following:

a. Connections to seismic force-resisting elements, excluding anchors (bolts or rods) embedded in concrete, shall be designed to develop Ω_0 times the calculated connection design force. ~~Connections and attachments for anchorage and other lateral force-resisting components shall be designed to develop the strength of the anchor (e.g., minimum published yield strength, F_y in direct tension, plastic bending moment), or Ω_0 times the calculated element design force.~~ For anchors (bolts or rods) embedded in concrete, the design of the anchor embedment shall meet the requirements of Section 15.7.5. Additionally, the connection of the anchors to the tank or vessel shall be designed to develop the lesser of the strength of the anchor in tension as determined by the reference document or Ω_0 times the calculated anchor design force. The overstrength requirements of Section 12.4.3, and the Ω_0 values tabulated in Table 15.4-2, do not apply to the design of walls, including interior walls, of tanks or vessels.

(no change to Items b. through f.)

15.7.5 Anchorage

Tanks and vessels at grade are permitted to be designed without anchorage where they meet the requirements for unanchored tanks in reference documents. Tanks and vessels supported above grade on structural towers or building structures shall be anchored to the supporting structure.

The following special detailing requirements shall apply to steel tank and vessel anchor bolts in SDC C, D, E, and F. ~~seismic regions where $S_{DS} > 0.5$, or where the structure is classified as Occupancy Category IV.~~ Anchorage shall be in accordance with Section 15.4.9, whereby the anchor embedment into the concrete shall be designed to develop the steel strength of the anchor in tension. The steel strength of the anchor in tension shall be determined in accordance with ACI 318, Appendix D, Eq. D-3. The anchor shall have a minimum gauge length of eight diameters. Post-installed anchors are permitted to be used in accordance with Section 15.4.9.3 provided the anchor embedment into the concrete shall be designed to develop the steel strength of the anchor in tension. In either case, the load combinations of Section 12.4.3 are not to be used to size the anchor bolts for tanks and horizontal and vertical vessels.

~~a. Hooked anchor bolts (L- or J-shaped embedded bolts) or other anchorage systems based soley on bond or mechanical friction shall not be used where $S_{DS} \geq 0.33$. Postinstalled anchors are permitted to be used provided that testing validates their ability to develop yield load in the anchor under cyclic loads in cracked concrete.~~

~~b. Where anchorage is required, the anchor embedment into the foundation shall be designed to develop the minimum specified yield strength of the anchor.~~

15.7.11.7 Supports and Attachments for Boilers and Pressure Vessels

Attachments to the pressure boundary and support for boilers and pressure vessels shall meet the following requirements:

a. Attachments and supports transferring seismic loads shall be constructed of ductile materials suitable for the intended application and environmental conditions.
b. ~~Seismic anchorages embedded in concrete shall be ductile and detailed for cyclic loads.~~ Anchorage shall be in accordance with Section 15.4.9, whereby the anchor embedment into the concrete is designed to develop the steel strength of the anchor in tension. The steel strength of the anchor in tension shall be determined in accordance with ACI 318 Appendix D Eq. D-3. The anchor shall have a minimum gauge length of eight diameters. The load combinations with overstrength of Section 12.4.3 are not to be used to size the anchor bolts for tanks, horizontal and vertical vessels.

(no change to balance of section)

Analysis and Commentary

The addition of Section 15.4.9 provides a consistent treatment of anchorage for nonbuilding structures throughout Chapter 15. It is more of a clarification. It also provides coordination with changes to Section 15.7 that are discussed here.

The changes to Section 15.7.3 are to make the intent of the provision clear. The intent is to design anchor attachments such that the anchor will yield before the anchor attachment to the structure fails. It also clarifies that the anchor rod embedment requirements are to be based on the requirements of Section 15.7.5, not 15.7.3(a). This again is to ensure that the anchor yields before the anchor attachment fails.

The changes to Sections 15.7.5 and 15.7.11.7 are to update the requirements for anchor bolts and anchor bolt embedment design to meet ACI 318 Appendix D. Anchor bolts used for tanks and vessels must yield under seismic loads to provide the required ductility. Replacing the vague requirement for designing the embedment for the minimum published yield strength of the anchor bolt with the terminology "tensile or shear strength of steel element," which is similar to that of ACI 318, ensures that anchor reinforcement is provided and that the anchor bolt will yield. Hooked anchor bolts are also deleted. Historically these bolts have performed poorly in seismic events. The Seismic Subcommittee of ASCE 7 decided to eliminate the use of these bolts.

Anchors in concrete used for nonbuilding structure anchorage in Chile
Courtesy of J. Silva, ASCE/SEI 2010 Chile Earthquake Reconnaissance Team

15.6.1 Modification

Earth-Retaining Structures

At a Glance

Conflicts and inaccuracies are removed from Section 15.6.1, Earth-Retaining Structures.

2010 Standard

15.6.1 Earth-Retaining Structures

This section applies to all earth-retaining structures assigned to Seismic Design Category D, E, or F. The lateral earth pressures due to earthquake ground motions shall be determined in accordance with Section 11.8.3 ~~for Seismic Design Categories B, C, D, E, and F with a geotechnical analysis prepared by a registered design professional~~.

The risk category shall be determined by the proximity of the earth-retaining structure ~~retaining wall~~ to other buildings and ~~other~~ structures. If failure of the earth-retaining structure ~~retaining wall~~ would affect the ~~an~~ adjacent building or structure, the risk category shall not be less than that of the adjacent building or structure~~, as determined in Table 1-1~~. Earth-retaining walls are permitted to be designed for seismic loads as either yielding or nonyielding walls. Cantilevered reinforced concrete or masonry retaining walls shall be assumed to be yielding walls and shall be designed as simple flexural wall elements.

Analysis and Commentary

The deletion of part of the second sentence of the first paragraph of ASCE 7-05 Section 15.6.1 eliminates a conflict with the first sentence, which limits the scope to Seismic Design Categories D, E, and F, and with the referenced Section 11.8.3, which is limited in applicability to Seismic Design Categories D, E, and F.

The reference to Table 1-1 is deleted because risk category is not determined in Table 1-1 and because one of the purposes of this section is to place a special condition on the determination of risk category not found elsewhere in the standard.

Requirements for earth-retaining structure are revised in Section 15.6.1 of ASCE 7-10.

15.6.5.1, 15.7.6.1.2 Modification

Freeboard, Height of Sloshing Wave

At a Glance

Equation 15.6-1 for minimum freeboard in Section 15.6, General Requirements for Nonbuilding Structures Not Similar to Buildings, and Equation 15.7-13 for the height of sloshing wave in Section 15.7.6, Ground-Supported Storage Tanks for Liquids, are modified.

2010 Standard

15.6.5.1 Freeboard

Sloshing of the liquid within the secondary containment area shall be considered in determining the height of the impound. Where the primary containment has not been designed with a reduction in the structure category (i.e., no reduction in importance factor I_e) as permitted by Section 1.5.3, no freeboard provision is required. Where the primary containment has been designed for a reduced structure category (i.e., importance factor I_e reduced) as permitted by Section 1.5.3, a minimum freeboard, δ_s, shall be provided where

$$\delta_s = \cancel{0.50}\ \underline{0.42} D S_{ac} \quad (15.6\text{-}1)$$

where S_{ac} is the spectral acceleration of the convective component and is determined according to the procedures of Section 15.7.6.1 using 0.5 percent damping. For circular impoundment dikes, D shall be taken as the diameter of the impoundment dike. For rectangular impoundment dikes, D shall be taken as the plan dimension of the impoundment dike, L, for the direction under consideration.

15.7.6.1.2 Sloshing Sloshing of the stored liquid shall be taken into account in the seismic design of tanks and vessels in accordance with the following requirements:

a. The height of the sloshing wave, δ_s, shall be computed using Eq. 15.7-13 as follows:

$$\delta_s = \cancel{0.50}\ \underline{0.42} D_i \underline{I_e} S_{ac} \quad (15.7\text{-}13)$$

For cylindrical tanks, D_i shall be the inside diameter of the tank; for rectangular tanks, the term D_i shall be replaced by the longitudinal plan dimension of the tank, L, for the direction under consideration.

(no change to remainder of section)

Analysis and Commentary

ASCE 7-05 and the 2005 edition of API 650 Appendix E introduced a new equation for determining the seismic wave height in liquid storage tanks for use in setting freeboard requirements. This change introduced significant additional height requirements on large storage tanks located in high seismic regions. While the introduction of the new equation provided a simplification of the previously applied equation and had minimal consequences for small tanks and tanks in low seismic areas, the application of this new equation to large tanks was unnecessarily conservative. The changes to Sections 15.6.5.1 and 15.7.6.1.2 make ASCE 7 consistent with recently revised tank standards such as API 620-08.

Comparisons of earlier equations to simplified equations $\delta_s = 0.42DS_{ac}$ and $\delta_s = 0.42DI_eS_{ac}$ have been made for various tank shapes. All first mode equations produced wave heights less than $\delta_s = 0.42DS_{ac}$. The ASCE and New Zealand multi-mode equations were found to produce values for some configurations a few percent higher than the $\delta_s = 0.42DS_{ac}$ value. However, the magnitude above the base equation was within the accuracy of the mode combination method.

Applying the $\delta_s = 0.50DS_{ac}$ equation adds 19% to the height of the wave and thus to the height of the required freeboard. The equation is inconsistent with all other international publications and the basis can be traced back to an approximation considered conservative. The equation is therefore changed to $\delta_s = 0.42DS_{ac}$.

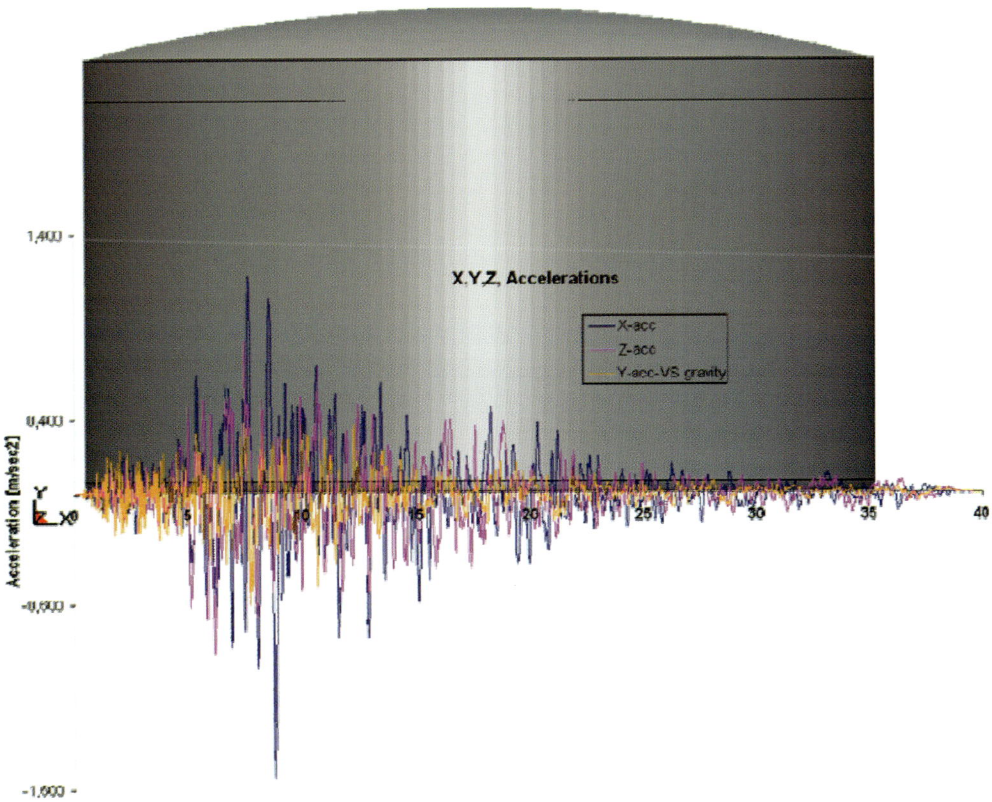

Design height of sloshing wave in stored liquid is reduced in ASCE 7-10.

Figure source: Dotoli, R., Lisi, D., Bardaro, D., Perillo, M., and Tomasi, M., "Sloshing Response of a LNG Storage Tank Subjected to Seismic Loading," Proceedings, 6th European LS-DYNA Users' Conference, May 2007, Gothenburg, Sweden.

15.7.6.1

Value of T_L in Design of Ground-Supported Tanks

Modification, Addition

At a Glance

Use of site-specific values of T_L that are less than mapped values are permitted in the design of ground-supported storage tanks for liquids, provided the site-specific values are no lower than certain specified floor values.

2010 Standard

15.7.6 Ground-Supported Storage Tanks for Liquids

15.7.6.1 General

(no change in section through Equation 15.7-11, then the following Exception is added)

EXCEPTION: For $T_c > 4$ s, S_{ac} is permitted be determined by a site-specific study using one or more of the following methods: (i) the procedures found in Chapter 21, provided such procedures, which rely on ground-motion attenuation equations for computing response spectra, cover the natural period band containing T_c, (ii) ground-motion simulation methods employing seismological models of fault rupture and wave propagation, and (iii) analysis of representative strong-motion accelerogram data with reliable long-period content extending to periods greater than T_c. Site-specific values of S_{ac} shall be based on one standard deviation determination. However, in no case shall the value of S_{ac} be taken as less than the value determined in accordance with Eq. 15.7-11 using 50% of the mapped value of T_L from Chapter 22.

The 80 percent limit on S_a required by Sections 21.3 and 21.4 shall not apply to the determination of site-specific values of S_{ac}, which satisfy the requirements of this exception. In determining the value of S_{ac}, the value of T_L shall not be less than 4 s.

Analysis and Commentary

Actual site-specific studies carried out since the introduction of the T_L requirements of ASCE 705 indicate that the mapped values of T_L are extremely conservative. Because a revision of the T_L maps is a time-consuming task that was not possible to complete during the ASCE 7-10 revision cycle, Section 15.7.6.1 is changed to effectively permit use of site-specific values of T_L that are less than the mapped values, with a floor of 4 seconds or one-half the mapped value of T_L. The exception is added under section15.7.6 because T_L is primarily used in tank design.

Use of site-specific values of T_L is now permitted in ASCE 7-10 for ground-supported tanks.

15.7.7.3, 23.1 Modification

Water Storage and Water Treatment Tanks and Vessels of Reinforced and Prestressed Concrete

At a Glance

Errors in Section 15.7.7.3 are corrected. Modifications are made to the three referenced standards, to make design requirements consistent with those of ASCE 7-10.

2010 Standard

15.7.7.3 Reinforced and Prestressed Concrete

Reinforced and prestressed concrete tanks shall be designed in accordance with the seismic requirements of AWWA D110, AWWA D115, or ACI 350.3 except that the importance factor, I_e, shall be determined according to Section 15.4.1.1 of this standard, the response modification coefficient, R, shall be taken from Table 15.4-2, and the design input forces for allowable stress design procedures shall be determined using the procedures of ACI 350.3 except ~~modified as follows~~

a. S_{ac} shall be substituted for C_c in ACI 350.3 Section 9.4.2 using Equations 15.7-10 for $T_c \leq T_L$ and 15.7-11 for $T_c > T_L$ from Section 15.7.6.1; and

b. The value of C_t from ACI 350.3 Section 9.4.3 shall be determined using the procedures of 15.7.2(c). The values of I, R_i, and b as defined in ACI 350.3 shall be taken as 1.0 in the determination of vertical seismic effects.

(ASCE 7-05 subsections a, b, and c are deleted)

23.1 CONSENSUS STANDARDS AND OTHER REFERENCE DOCUMENTS

ACI 350.3
Sections 15.7.6.1.1, 15.7.7.3
Standard Practice for the Seismic Design of Liquid-Containing Concrete Structures, 2006~~2001~~

AWWA D110
Section 15.7.7.3
Wire- and Strand-Wound Circular Prestressed Concrete Water Tanks, 2004~~1995~~

AWWA D115
Section 15.7.7.3
Tendon-Prestressed Concrete Water Tanks, ~~*Circular Pre-stressed Concrete Tanks with Circumferential Tendons*~~, 2006~~1995~~

Analysis and Commentary

The changes to Section 15.7.7.3 correct errors in ASCE 7-05 and update the three referenced standards.

The design requirements for water treatment tanks of concrete are found in Section 15.7.7.3.

Courtesy: Peripitus, commons.wikimedia.org, under the terms of GNU Documentation License, Version 1.2 or later

15.7.8.1, 15.7.13.1, 23.1 — Modification

Welded Steel Petrochemical and Industrial Tanks and Vessels Storing Liquids, Refrigerated Gas Liquid Storage Tanks and Vessels

At a Glance

Provisions for welded steel petrochemical and industrial tanks and vessels storing liquids are updated and the referenced standard is updated to the 2009 edition. Tanks and facilities for the storage of liquefied hydrocarbons and refrigerated liquids are now included within the scope of ASCE 7-10.

2010 Standard

15.7.8 Petrochemical and Industrial Tanks and Vessels Storing Liquids

15.7.8.1 Welded Steel

Welded steel petrochemical and industrial tanks and vessels storing liquids under an internal pressure of less than or equal 2.5 psig (17.2 kPa g) shall be designed in accordance with the seismic requirements of API 650. Welded steel petrochemical and industrial tanks and vessels storing liquids under an internal pressure of greater than 2.5 psig (17.2 kPa g) and less than or equal 15 psig (104.4 kPa g) shall be designed in accordance with the seismic requirements of API 620.~~Tanks designed in accordance with the requirements of API 620 shall also be designed in accordance with the seismic requirements of API 650.~~

15.7.13 Refrigerated Gas Liquid Storage Tanks and Vessels

15.7.13.1 General

Tanks and facilities for the storage of liquefied hydrocarbons and refrigerated liquids shall meet the requirements of this standard. ~~The seismic design of the tanks and facilities for the storage of liquefied hydrocarbons and refrigerated liquids is beyond the scope of this section. The design of such tanks is addressed in part by various reference documents listed in Chapter 23.~~ ~~**EXCEPTION:**~~ Low-pressure, welded steel storage tanks for liquefied hydrocarbon gas (e.g., LPG, butane, etc.) and refrigerated liquids (e.g., ammonia) shall be designed in accordance with the requirements of Section 15.7.8 and API 620.

23.1 CONSENSUS STANDARDS AND OTHER REFERENCE DOCUMENTS

API 620

Sections 15.4.1, 15.7.8.1, 15.7.13.1

Design and Construction of Large, Welded, Low Pressure Storage Tanks, 11~~10~~th edition, 2009~~2002~~

Analysis and Commentary

Previous editions of API 620 contained seismic provisions that were out-of-date and not aligned with the requirements of ASCE 7-05. The seismic provisions in the latest edition of API 620 Appendix L have been brought up to the requirements of ASCE 7-05 and API 650 Appendix E. In addition, special seismic requirements, more conservative than those of ASCE 7-05 and API 650, have been added to deal with the design of certain low-temperature and cryogenic storage tanks where other industry standards require two levels of ground motions.

Section 15.7.13.1 was modified to require that the seismic design of tanks and facilities for the storage of liquefied hydrocarbons and refrigerated liquids meet the provisions of ASCE 7 as a minimum.

Updated API 620-2009 is referenced for low-pressure welded steel storage tanks for liquefied hydrocarbon gas and refrigerated liquids in ASCE 7-10.

Courtesy: Cjp24, commons.wikimedia.org, under the terms of GNU Free Documentation License, Version 1.2 or later

Part VI
Chapter 16 Seismic Response History Procedures

16.1.3.2 Modification, Addition

Three-Dimensional Seismic Response History Analysis

At a Glance

The requirements in Chapter 16 for analysis using two components of horizontal ground motion (explicit three-dimensional seismic response history analysis) are modified to account for ground motions based on the maximum direction of response.

2010 Standard

16.1.3.2 Three-Dimensional Analysis

Where ~~3-D~~ three-dimensional ~~analysis is~~ analyses are performed, ground motions shall consist of pairs of appropriate horizontal ground motion acceleration components that shall be selected and scaled from individual recorded events. Appropriate ground motions shall be selected from events having magnitudes, fault distance, and source mechanisms that are consistent with those that control the maximum considered earthquake. Where the required number of recorded ground motion pairs ~~are~~is not available, appropriate simulated ground motion pairs ~~shall~~ are permitted to be used to make up the total number required. For each pair of horizontal ground motion components, a square root of the sum of the squares (SRSS) spectrum shall be constructed by taking the SRSS of the 5 percent-damped response spectra for the scaled components (where an identical scale factor is applied to both components of a pair). Each pair of motions shall be scaled such that ~~for each period between~~ in the period range from $0.2T$ and $1.5T$, the average of the SRSS spectra from all horizontal component pairs does not fall below~~1.3 times and~~ the corresponding ordinate of the ~~design~~ response spectrum used in the design, determined in accordance with Section 11.4.5 or ~~21.2~~11.4.7~~, by more than 10 percent.~~

At sites within 3 miles (5 km) of the active fault that controls the hazard, each pair of components shall be rotated to the fault-normal and fault-parallel directions of the causative fault and shall be scaled so that the average of the fault-normal components is not less than the MCE_R response spectrum for the period range from $0.2T$ to $1.5T$.

Analysis and Commentary

The revised language specifically changes the requirements in Chapter 16 (Seismic Response History Procedures) for analysis using two components of horizontal ground motion (e.g., for explicit three-dimensional analysis) to account for ground motions based on the maximum direction of response. Language in the last sentence of the only paragraph in Section 16.1.3.2 of ASCE 7-05 and now the first paragraph in the same section of ASE 7-10 has been simplified in view of the following:

$$\begin{aligned}\text{SRSS spectrum} &\geq 1.3 \times 0.9 \text{ geomean design spectrum (ASCE 7-05)}\\ &= 1.17 \text{ geomean design spectrum}\\ &\approx 1.0 \text{ maximum-direction design spectrum (ASCE 7-10)}\end{aligned}$$

For sites within approximately three miles of an active fault that controls the ground-motion hazard, the near-field strong-motion database indicates that the fault-normal (FN) direction is (or close to) the direction of maximum ground motion for periods around 1.0 sec and greater (Huang et al., 2008; Watson-Lamprey and Boore, 2007). In this case, the two horizontal components of a selected record should be transformed so that one component is the motion in the FN direction and the other component is the motion in the fault-parallel (FP) direction. Scaling so that the average FN component response spectrum is at the level of the MCE response spectrum ensures that the FN components will not be underestimated, which would happen if the SRSS rule were applied at short distances. The same scale factor selected for the FN component of a given record is to be used for the FP component as well.

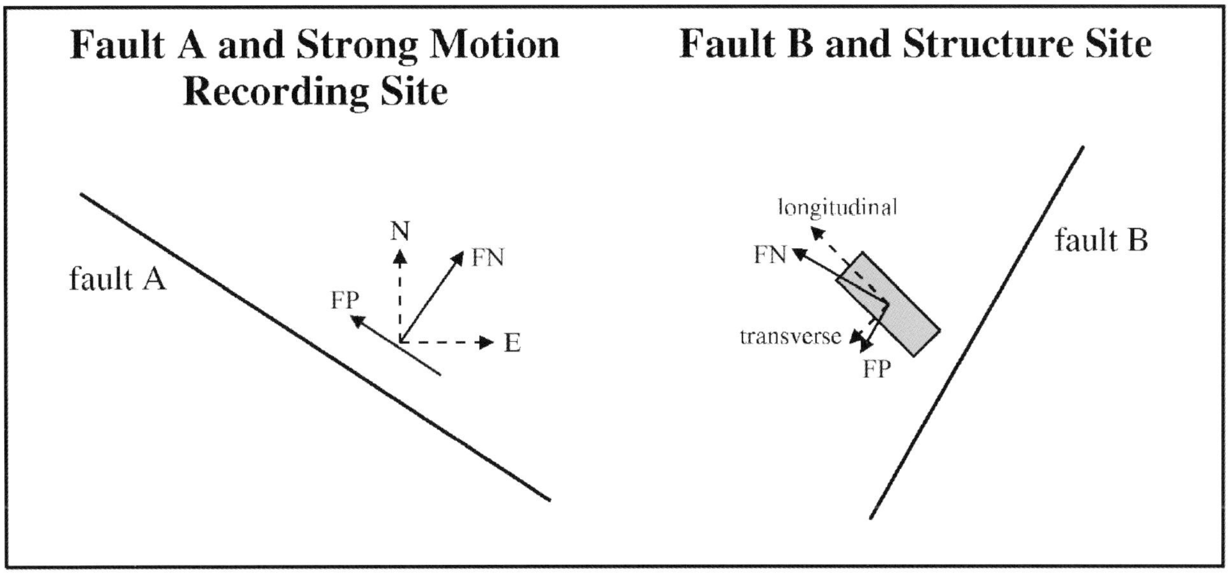

Recording (left) and application (right) of fault-normal and fault-parallel components of ground motion

16.1.4 Modification

Response Parameters in Linear Response History Analysis

At a Glance

Force-related response quantities resulting from linear response history analysis shall be multiplied by I_e/R, while displacement-related response quantities shall be multiplied by C_d/R.

2010 Standard

16.1.4 Response Parameters

For each ground motion analyzed, the individual response parameters shall be multiplied by the following scalar quantities: ~~scalar quantity I/R where I is the importance factor determined in accordance with Section 11.5.1 and R is the response modification coefficient selected in accordance with Section 12.2.1.~~

a. Force response parameters shall be multiplied by I_e/R, where I_e is the importance factor determined in accordance with Section 11.5.1 and R is the Response Modification Coefficient selected in accordance with Section 12.2.1.

b. Drift quantities shall be multiplied by C_d/R, where C_d is the deflection amplification factor specified in Table 12.2-1.

For each ground motion i, where i is the designation assigned to each ground motion, the maximum value of the base shear, V_i, member forces, Q_{Ei}, scaled as indicated in the preceding text and story drifts, Δ_i, at each story~~,~~ as defined in Section 12.8.6 ~~scaled as indicated in the preceding text~~ shall be determined.

(no changes in remainder of Section 16.1.4.)

Analysis and Commentary

In modifying the results of linear response history analysis to account for permitted inelastic response, ASCE 7-05 made no distinction between force response parameters and drift quantities. For consistency with all other analysis procedures, a distinction is now made in ASCE 7-10. Inelastic drifts should be determined as elastic drifts from linear response history analysis amplified by C_d/R, while inelastic force parameters are to be determined as elastic force parameters from linear response history analysis multiplied by I_e/R.

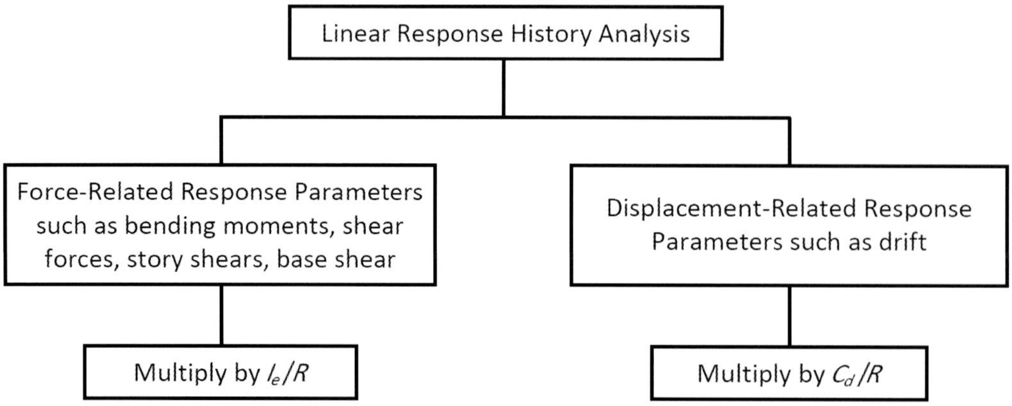

16.1.5 Addition

Horizontal Shear Distribution in Linear Response History Analysis

At a Glance

Consideration of torsion is made consistent between linear response history analysis and modal response spectrum analysis.

2010 Standard

16.1.5 Horizontal Shear Distribution

The distribution of horizontal shear shall be in accordance with Section 12.8.4 except that amplification of torsion in accordance with Section 12.8.4.3 is not required where accidental torsion effects are included in the dynamic analysis model.

Analysis and Commentary

This change makes consideration of accidental torsion for linear response history analysis (Section 16.1.5) consistent with that for modal response spectrum analysis (Section 12.9.5). Section 12.8.4 requires that the seismic design story shear, Vx, be distributed to the various vertical elements of the seismic force-resisting system in the story under consideration based on the relative lateral stiffness of the vertical resisting elements and the diaphragm.

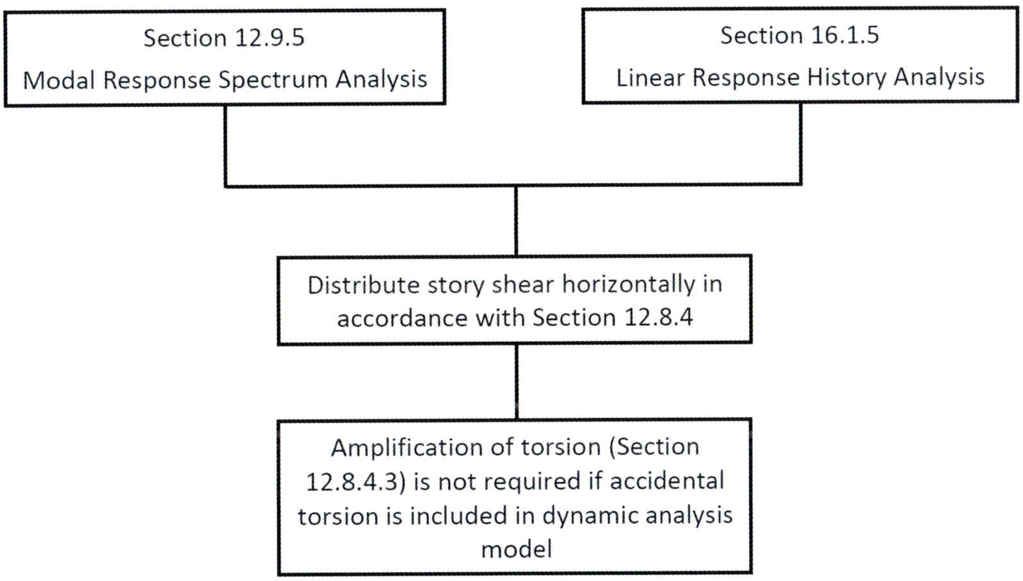

Part VII
Chapter 19 Soil-Structure Interaction for Seismic Design

Table 19.2-1

Modification

Values of Shear Wave Velocity and Shear Modulus for Soil-Structure Interaction Analysis

At a Glance

Values of shear wave velocity and shear modulus are now made functions of site class.

2010 Standard

~~TABLE 19.2-1 VALUES OF G/G_O AND v_s/v_{so}~~

	~~Spectral Response Acceleration, S_{D1}~~			
	~~$\leq$0.10~~	~~$\leq$0.15~~	~~$\leq$0.20~~	~~$\geq$0.30~~
~~Value of G/G_O~~	~~0.81~~	~~0.64~~	~~0.49~~	~~0.42~~
~~Value of v_s/v_{so}~~	~~0.9~~	~~0.8~~	~~0.7~~	~~0.65~~

Table 19.2-1 Values of G/G_o and v_s/v_{so}

	Value of v_s/v_{so}			Value of G/G_o		
	$S_{DS}/2.5$			$S_{DS}/2.5$		
Site Class	$\leq$ 0.1	0.4	$\geq$ 0.8	$\leq$ 0.1	0.4	$\geq$ 0.8
A	1.00	1.00	1.00	1.00	1.00	1.00
B	1.00	0.97	0.95	1.00	0.95	0.90
C	0.97	0.87	0.77	0.95	0.75	0.60
D	0.95	0.71	0.32	0.90	0.50	0.10
E	0.77	0.22	[a]	0.60	0.05	[a]
F	[a]	[a]	[a]	[a]	[a]	[a]

Note: Use straight-line interpolation for intermediate values of $S_{DS}/2.5$.

[a] Should be evaluated from site specific analysis

Analysis and Commentary

The ASCE 7-05 table for values of G/G_O and v_s/v_{so} used single values of modulus reduction factors, which did not account for differences in shear strain associated with soils having different stiffnesses. The revised Table 19.2-1 was developed for FEMA 356, *Prestandard and Commentary for the Seismic Rehabilitation of Buildings*, to correct that error. The correction was made to the 2009 NEHRP *Provisions* and now has been made in ASCE 7-10 as well.

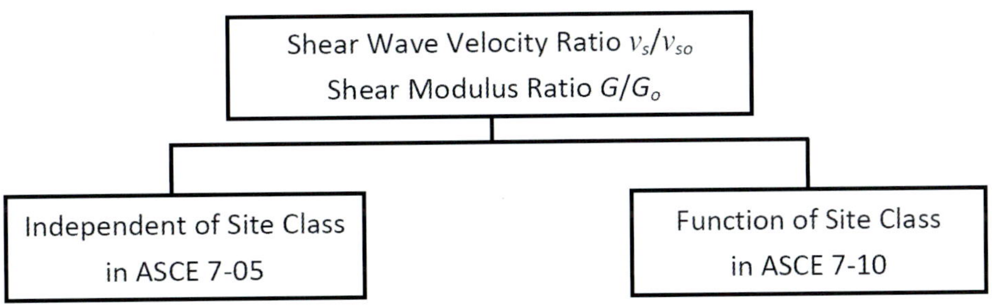

Figure 19.2-1 Modification

Foundation Damping Factor in Soil-Structure Interaction Analysis

At a Glance

ASCE 7-05 Figure 19.2-1, Foundation Damping Factor, has been replaced with a new figure.

2010 Standard

ASCE 7-05 Figure 19.2-1 is replaced with the following figure:

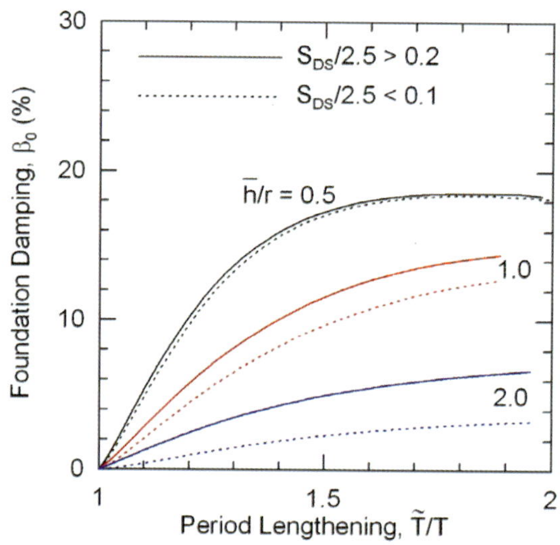

Figure 19.2-1 Foundation Damping Factor

Analysis and Commentary

ASCE 7-05 Figure 19.2-1 could not be reproduced from source articles used to derive it. To remedy this situation, the figure shown above was developed in the ATC-55 project and included in the resulting FEMA 440 report, *Improvement of Nonlinear Static Seismic Analysis Procedures*. The figure was incorporated in the 2009 NEHRP *Provisions* and is now included in ASCE 7-10 as well.

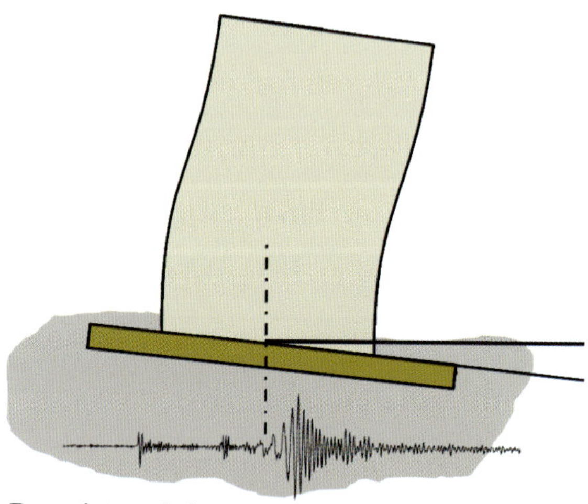

Damping and shear modulus of soil influence dymamic response of structure

Part VIII
Chapter 21 Site-Specific Ground Motion Procedures for Seismic Design

Figure 21.2-1

Deterministic Lower Limit on MCE_R Response Spectrum from Site Response Analysis

Modification, Correction

At a Glance

ASCE 7-05 Figure 21.2-1, which was incorrect, has been replaced with a corrected figure.

2010 Standard

ASCE 7-05 Figure 21.2-1 is replaced with the following figure:

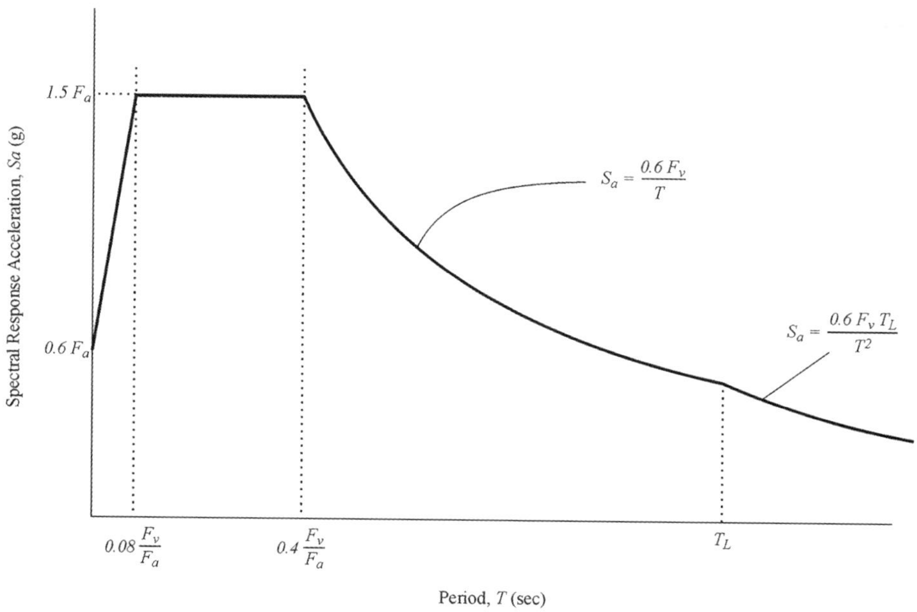

FIGURE 21.2-1 Deterministic Lower Limit on MCE_R Response Spectrum

Analysis and Commentary

Figure 21.2-1 in ASCE 7-05 did not show the "ramp" building up to the "flat top" or the segment beyond the long-period transition period. This change corrects those omissions with the above replacement figure.

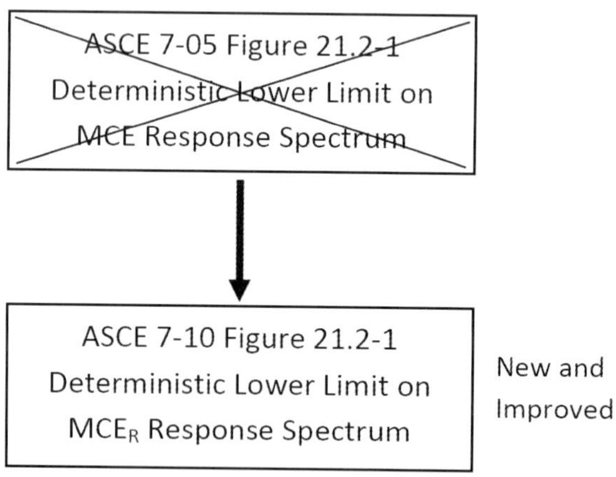

21.4

Design Acceleration Parameters from Site-Specific Ground Motion Procedures

Modification, Addition

At a Glance

Code language is added to clarify that the values of S_{DS} and S_{DI} given by Section 21.4 are not to be used in the determination of the ELF base shear by Section 12.8.

2010 Standard

21.4 DESIGN ACCELERATION PARAMETERS

Where the site-specific procedure is used to determine the design ground motion in accordance with Section 21.3, the parameter S_{DS} shall be taken as the spectral acceleration, S_a, obtained from the site-specific spectra at a period of 0.2 s, except that it shall not be taken less than 90 percent of the peak spectral acceleration, S_a, at any period larger than 0.2 s. The parameter S_{DI} shall be taken as the greater of the spectral acceleration, S_a, at a period of 1 s or two times the spectral acceleration, S_a, at a period of 2 s. The parameters S_{MS} and S_{M1} shall be taken as 1.5 times S_{DS} and S_{D1}, respectively. The values so obtained shall not be less than 80 percent of the values determined in accordance with Section 11.4.3 for S_{MS} and S_{M1} and Section 11.4.4 for S_{DS} and S_{D1}.

<u>For use with the Equivalent Lateral Force Procedure, the site specific spectral acceleration, S_a, at T shall be permitted to replace S_{D1}/T in Equation 12.8-3 and $S_{D1}T_L/T^2$ in Equation 12.8-4. The parameter S_{DS} calculated per this section shall be permitted to be used in Equations 12.8-2 and 12.8-5, 15.4-1, and 15.4-3. The mapped value of S_1 shall be used in Equation 12.8-6, 15.4-2, and 15.4-4.</u>

Analysis and Commentary

Section 21.4 specifies the approach to determining design acceleration parameters S_{DS} and S_{D1} when the site-specific procedure is used. The values of S_{DS} and S_{D1} are important in the determination of the following:

1. Seismic Design Category (S_{DS} and S_{D1})
2. Load combinations (S_{DS})
3. Out of plane wall and anchorage forces (S_{DS})
4. C_u for upper bound on rationally computed period (S_{D1})
5. Nonstructural design force (S_{DS})
6. Scaling in modal response spectrum analysis (refers to 85% of value given by ELF formulas using both S_{DS} and S_{D1})

However, the intent was not to use the values of S_{DS} and S_{D1} given by Section 21.4 in the determination of the ELF base shear by Section 12.8. Rather, the site-specific spectrum developed utilizing Chapter 21 should be used for the latter purpose. This change clarifies this intent by specifying the appropriate modifications to Equations 12.8-3 and 12.8-4 when utilizing the site-specific spectrum approach. The change further clarifies that the parameter S_{DS} is permitted to be used in Equations 12.8-2 and 12.8-5 (also 15.4-1 and 15.4-3 for nonbuilding structures) and that the mapped value of S_1 is to be used in Equation 12.8-6 (and also 15.4-2 and 15.4-4 for nonbuilding structures).

Equation No.	Equation	Use
12.8-2	$C_s = S_{DS}/(R/I_e)$	S_{DS} from 21.4*
12.8-3	$C_s = S_{D1}/\ T(R/I_e)$ for $T \leq T_L$	S_a at T, not S_{D1}/T
12.8-4	$C_s = S_{D1}T_L /\ T^2(R/I_e)$ for $T > T_L$	S_a at T, not $S_{D1}T_L/T^2$
12.8-5	$C_s = 0.044 S_{DS} I_e \geq 0.01$	S_{DS} from 21.4*
12.8-6	$C_s = 0.5 S_1 /(R/I_e)$	Mapped value of S_1
15.4-1	$C_s = 0.044 S_{DS} I_e \geq 0.03$	S_{DS} from 21.4*
15.4-2	$C_s = 0.8 S_1 /(R/I_e)$	Mapped value of S_1
15.4-3	$C_s = 0.044 S_{DS} I_e \geq 0.01$	S_{DS} from 21.4*
15.4-4	$C_s = 0.5 S_1 /(R/I_e)$	Mapped value of S_1

* Permitted, not mandatory

ABOUT THE AUTHORS

S. K. Ghosh, Ph.D. is President of S. K. Ghosh Associates Inc., a seismic and building code consulting firm based in Palatine, IL and Aliso Viejo, CA. He has influenced seismic design provisions in the United States for many years by authoring many publications and by serving on or chairing numerous committees and advisory panels.

Susan Dowty, S.E. is Vice President of S. K. Ghosh Associates Inc. She has over twenty years of experience in the administration and development of building code provisions. She served as the consultant for the reformat of the ASCE 7-05 seismic provisions and served on the California Building Standards Commission 2005-2009.

Prabuddha Dasgupta, Ph.D., P.E. is Engineering Manager of S. K. Ghosh Associates Inc. He is actively engaged in interpretation of building code provisions and in the development and publication of code support literature and software.

Notes